Workbook
Progress in Mathematics

SADLIER-OXFORD

Rose Anita McDonnell

Catherine D. LeTourneau

Anne Veronica Burrows

M. Winifred Kelly

Colleen A. Dougherty

Mary Grace Fertal

Helen T. Smythe

Monica T. Sicilia

with
Dr. Elinor R. Ford

Sadlier-Oxford
A Division of William H. Sadlier, Inc.

Contents

Name _____

Write how many.

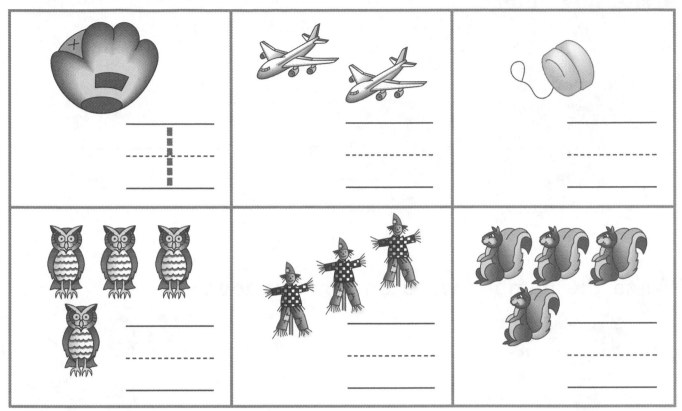

Write the number word and the number.

Write how many.

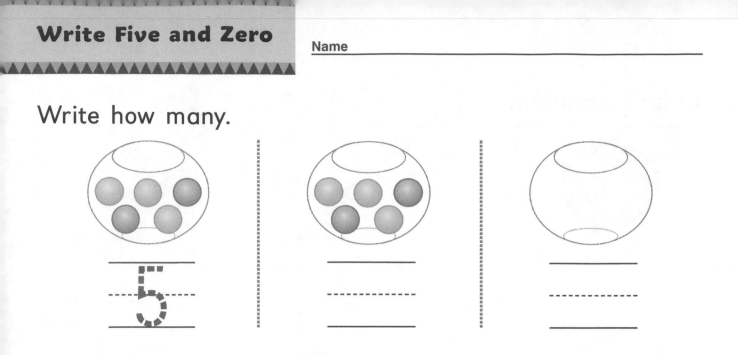

_____5_____   _____   _____

Write the number word and the number.

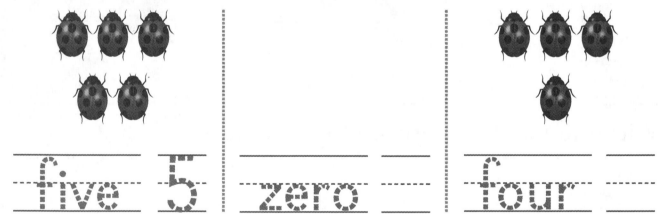

five 5   zero   four

Write the number word and the number.
Draw ● for each number.

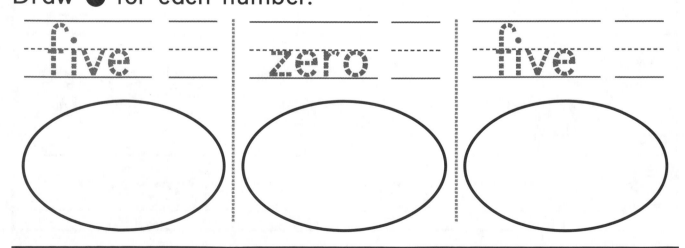

five   zero   five

Use with Lesson 1-3, text pages 23–24.

Draw one more. Write the number.

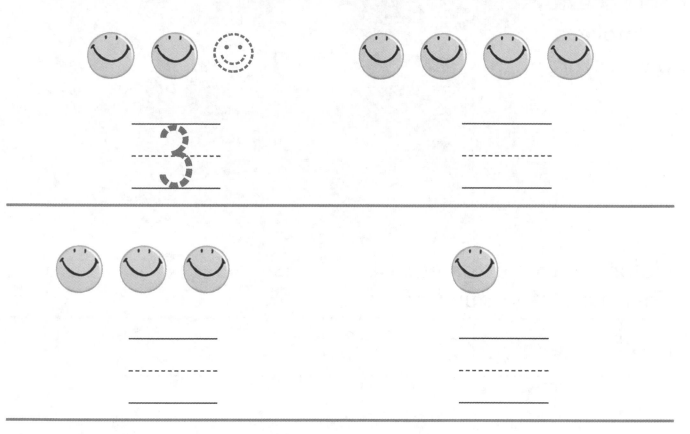

✗ to show one fewer. Write the number.

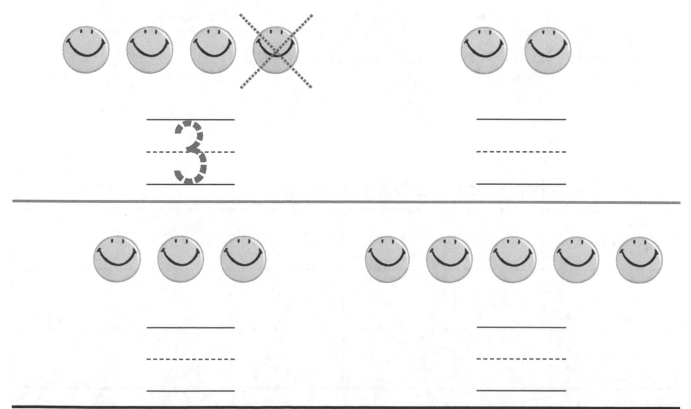

Name _____

Sort the toys to make a picture graph.

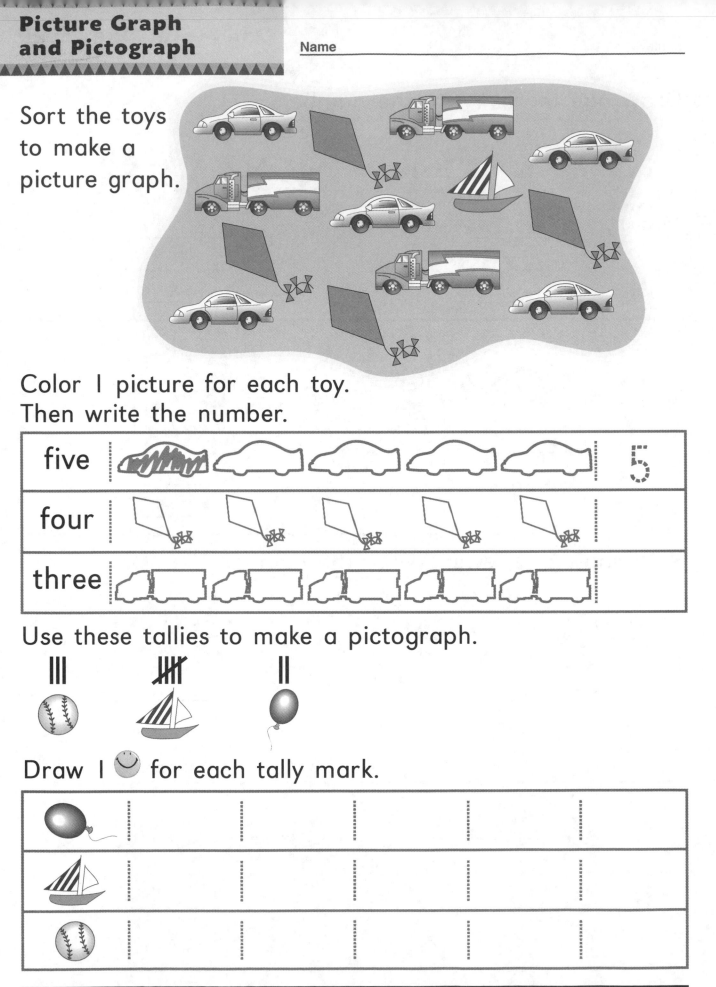

Color 1 picture for each toy.
Then write the number.

five						5
four						
three						

Use these tallies to make a pictograph.

||| ||||| ||

Draw 1 😊 for each tally mark.

4   **Use with Lesson 1-5, text pages 27–28.**

Write how many.

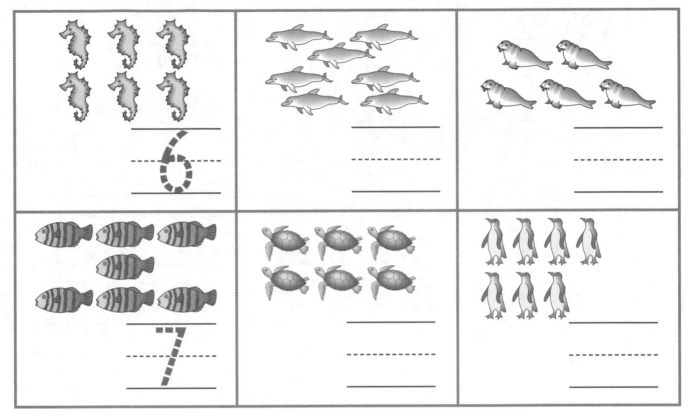

Write the number word and the number.

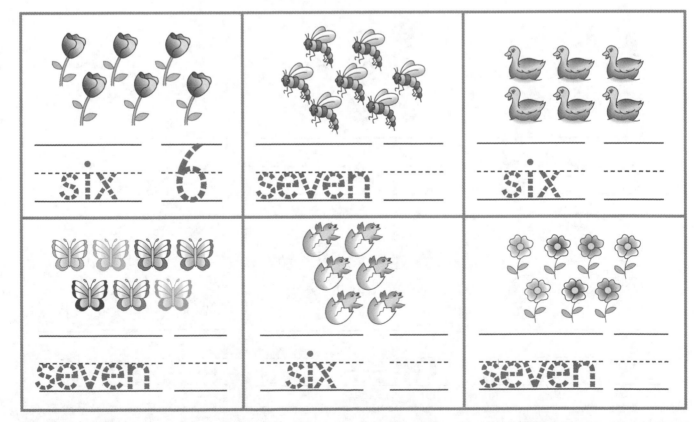

Write Eight and Nine

Write how many.

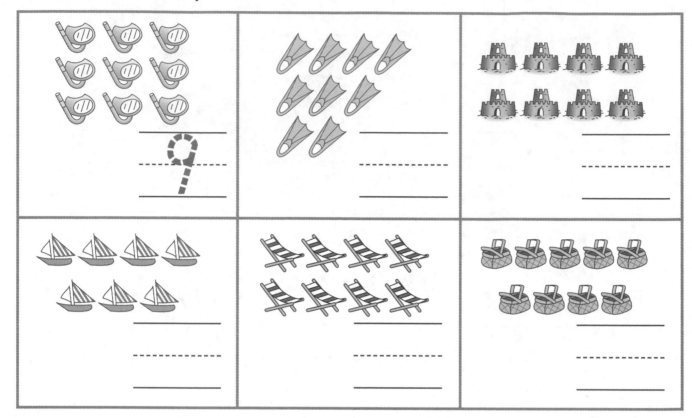

Write the number word and the number.

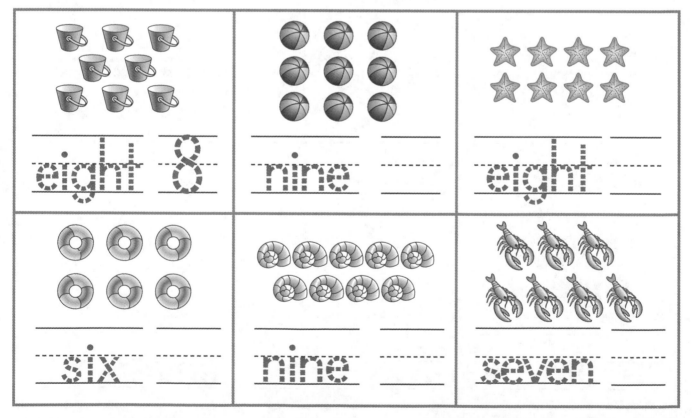

6

Name _____

Write how many.

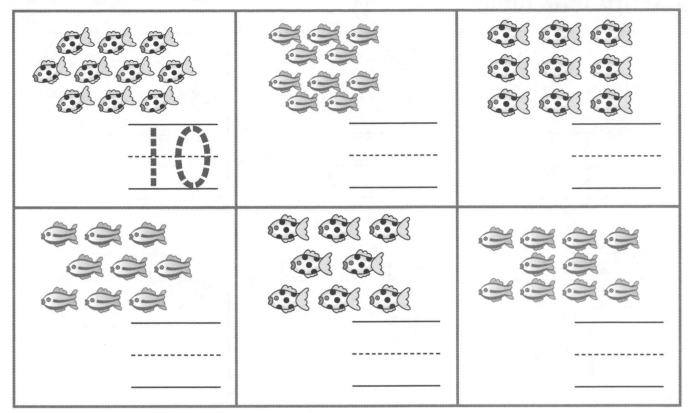

Write the number word and the number.

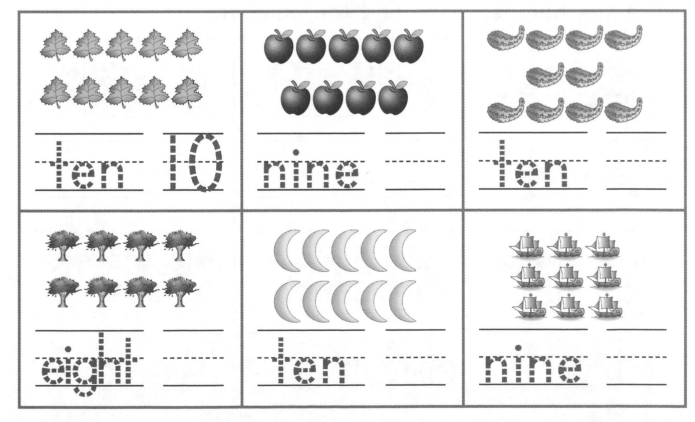

Use with Lesson I-8, text pages 33–34.

Write how many.

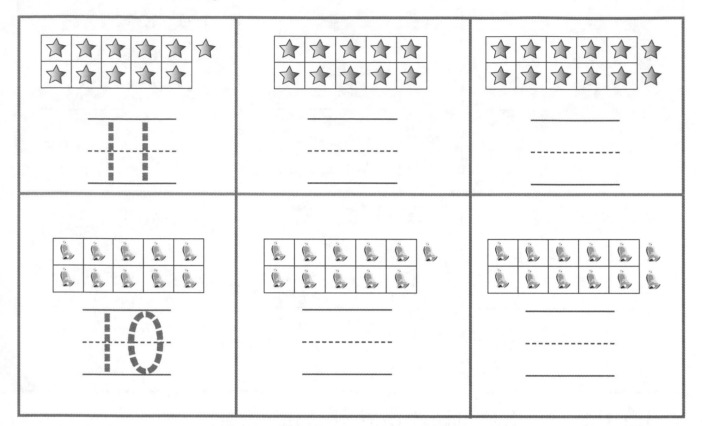

Write the number word and the number.

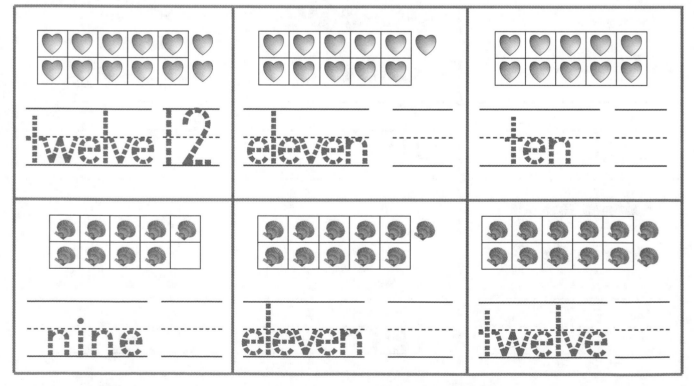

Use with Lesson I-9, text pages 35–36.

Trace the number.
Draw to show the number.

zero	0	
one	1	○
two	2	○○
three	3	
four	4	
five	5	
six	6	
seven	7	
eight	8	
nine	9	
ten	10	
eleven	11	
twelve	12	

Use with Lesson I-10, text pages 37–38.

Name _____

Write the missing numbers.

1. 0, 1, __2__, 3, ____, 5, ____, 7, ____, 9, ____, 11, 12

2. 0, ____, 2, 3, ____, 5, 6, ____, 8, 9, 10, ____, 12

Write how many.

3. __5__   ____   ____   ____

Now order the numbers. ____ ____ ____ ____

Write how many.

4. ____   ____   ____   ____

Now order the numbers. ____ ____ ____ ____

Write the missing numbers.

5. 4, __5__, 6, ____   6. ____, 9, 10, ____

7. ____, 3, ____, 5   8. 6, ____, ____, 9

9. ____, 10, 11, ____   10. ____, ____, 2, 3

Name _____

Write the missing numbers.

1.

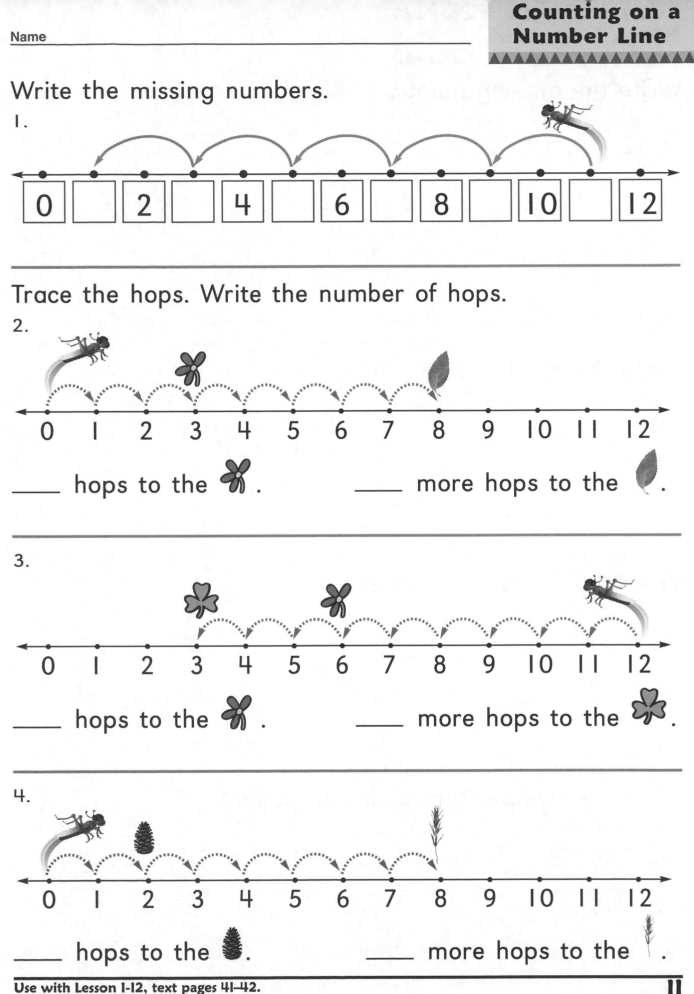

| 0 | | 2 | | 4 | | 6 | | 8 | | 10 | | 12 |

Trace the hops. Write the number of hops.

2.

0 1 2 3 4 5 6 7 8 9 10 11 12

____ hops to the ❀ .   ____ more hops to the 🍃 .

3.

0 1 2 3 4 5 6 7 8 9 10 11 12

____ hops to the ❀ .   ____ more hops to the ☘ .

4.

0 1 2 3 4 5 6 7 8 9 10 11 12

____ hops to the 🌰 .   ____ more hops to the 🌾 .

Counting Back; Before, After, Between

Name _____

Write the missing numbers.

1. 12, 11, 10, _9_, ___, 7, 6, ___, 4, ___, 2, ___, 0

2. 12, 11, ___, ___, 8, 7, ___, ___, 4, 3, 2, 1, ___

3. 12, ___, 10, 9, ___, 7, ___, 5, 4, ___, ___, 1, 0

4. 12, 11, 10, ___, 8, ___, 6, ___, ___, 3, 2, ___, 0

Write the number just before.

5. _4_, 5   ___, 3   ___, 8

6. ___, 2   ___, 12   ___, 10

7. ___, 6   ___, 1   ___, 7

Write the number just after.

8. 8, _9_   1, ___   3, ___

9. 5, ___   10, ___   0, ___

10. 11, ___   4, ___   6, ___

Write the number that comes between.

11. 5, _6_, 7   1, ___, 3   8, ___, 10

12. 0, ___, 2   6, ___, 8   3, ___, 5

13. 7, ___, 9   2, ___, 4   10, ___, 12

Use with Lessons 1-13 and 1-14, text pages 43–44.

Name _____

Draw ⭕ to compare.
Write the missing numbers.

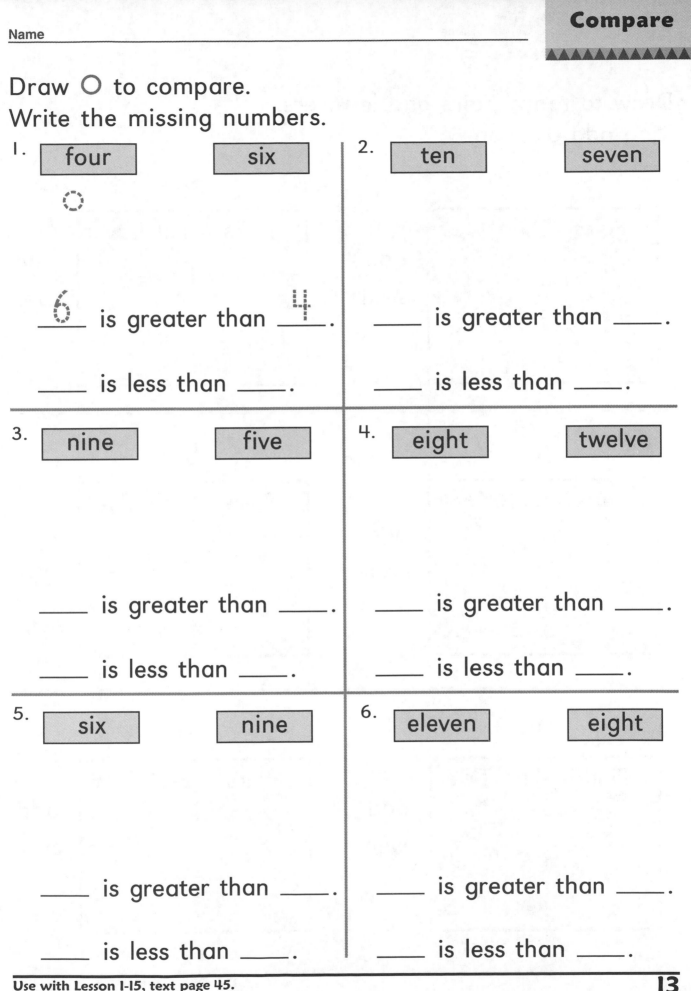

1. | four | six |

___6___ is greater than ___4___.

_____ is less than _____.

2. | ten | seven |

_____ is greater than _____.

_____ is less than _____.

3. | nine | five |

_____ is greater than _____.

_____ is less than _____.

4. | eight | twelve |

_____ is greater than _____.

_____ is less than _____.

5. | six | nine |

_____ is greater than _____.

_____ is less than _____.

6. | eleven | eight |

_____ is greater than _____.

_____ is less than _____.

Use with Lesson 1-15, text page 45.

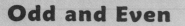

Draw to make pairs and leftovers.
Ring odd or even.

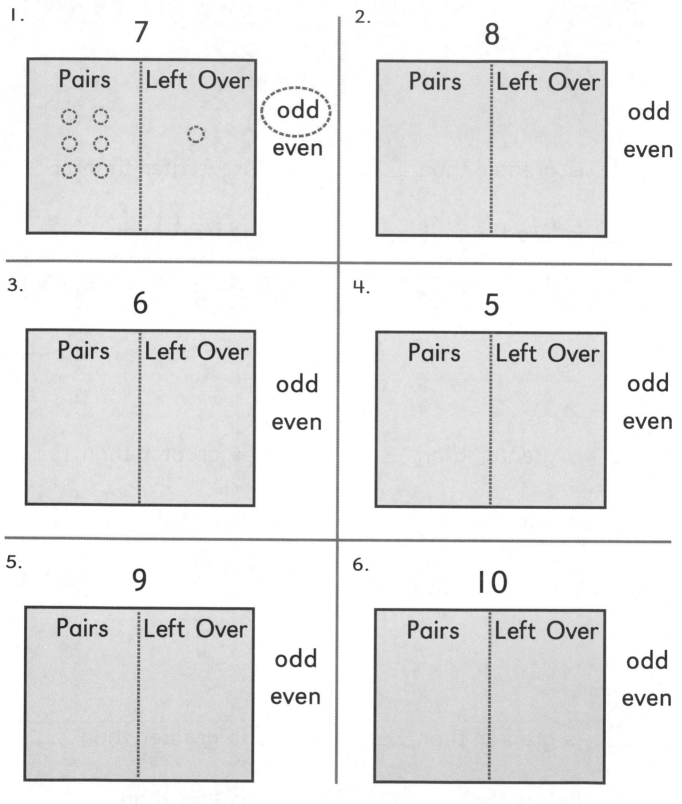

1.

7

Pairs	Left Over
○ ○	
○ ○	○
○ ○	

(odd)
even

2.

8

Pairs	Left Over

odd
even

3.

6

Pairs	Left Over

odd
even

4.

5

Pairs	Left Over

odd
even

5.

9

Pairs	Left Over

odd
even

6.

10

Pairs	Left Over

odd
even

Use with Lesson 1-16, text page 46.

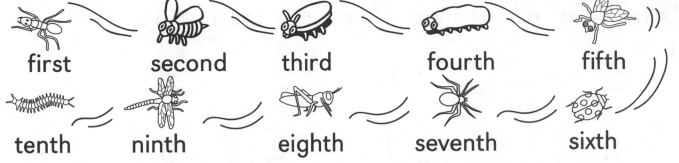

first   second   third   fourth   fifth

tenth   ninth   eighth   seventh   sixth

Ring the position of each bug.

		(second)	seventh	third
		second	fifth	third
		fourth	first	tenth
		ninth	eighth	fourth
		second	tenth	sixth

Color the:

tenth box. ☐ ☐ ☐ ☐ ☐ ☐ ☐ ☐ ☐ ▨

eighth box. ☐ ☐ ☐ ☐ ☐ ☐ ☐ ☐ ☐ ☐

seventh box. ☐ ☐ ☐ ☐ ☐ ☐ ☐ ☐ ☐ ☐

ninth box. ☐ ☐ ☐ ☐ ☐ ☐ ☐ ☐ ☐ ☐

sixth box. ☐ ☐ ☐ ☐ ☐ ☐ ☐ ☐ ☐ ☐

Draw what comes next. Then write the number.

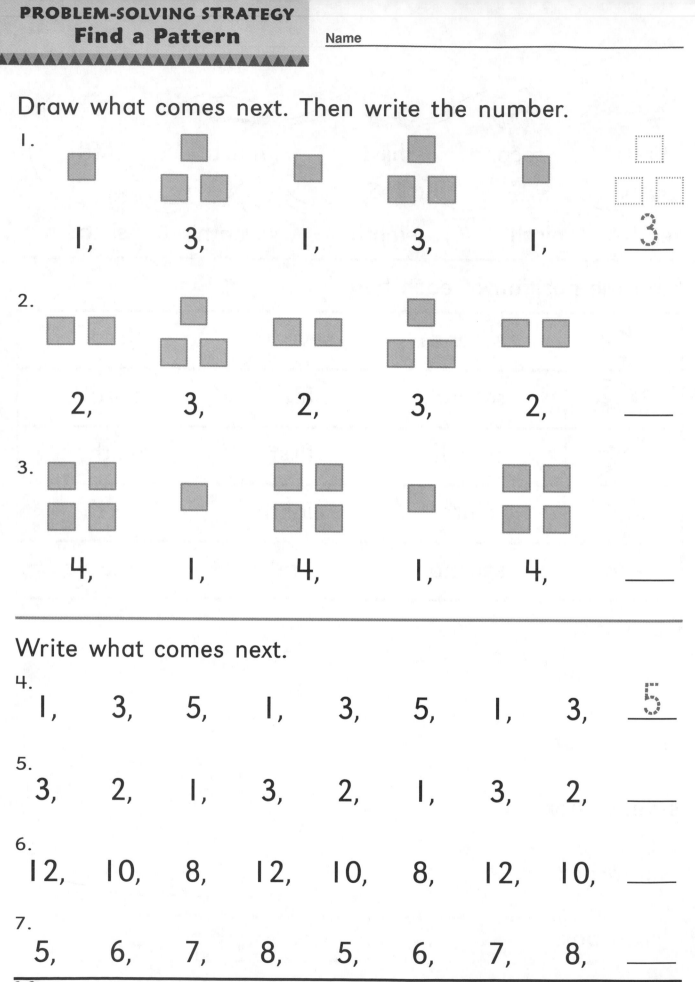

1. 1,   3,   1,   3,   1,   _3_

2. 2,   3,   2,   3,   2,   ____

3. 4,   1,   4,   1,   4,   ____

Write what comes next.

4. 1,   3,   5,   1,   3,   5,   1,   3,   _5_

5. 3,   2,   1,   3,   2,   1,   3,   2,   ____

6. 12,   10,   8,   12,   10,   8,   12,   10,   ____

7. 5,   6,   7,   8,   5,   6,   7,   8,   ____

Use with Lesson 1-18, text pages 49–50.

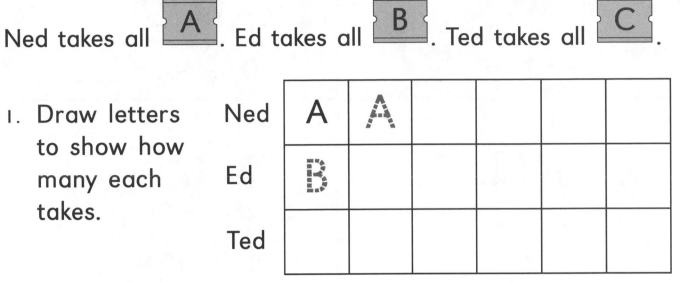

A A B B B C C C C C

Ned takes all A . Ed takes all B . Ted takes all C .

1. Draw letters to show how many each takes.

Ned	A	A			
Ed	B				
Ted					

Use the graph above. Ring the answer.

2. Who has more?   Ned   Ed

3. Who has fewer?   Ed   Ted

4. Ned and Ed together have

       more than Ted   as many as Ted

5. Ed gives 1 away.
   Now Ed has as many as   Ned   Ted

6. Ned gets more.
   Now Ned has as many as Ted.
   Now Ned has   2   3   5

7. How many will Ted have
   if he loses one?   2   4   6

Add.

1.

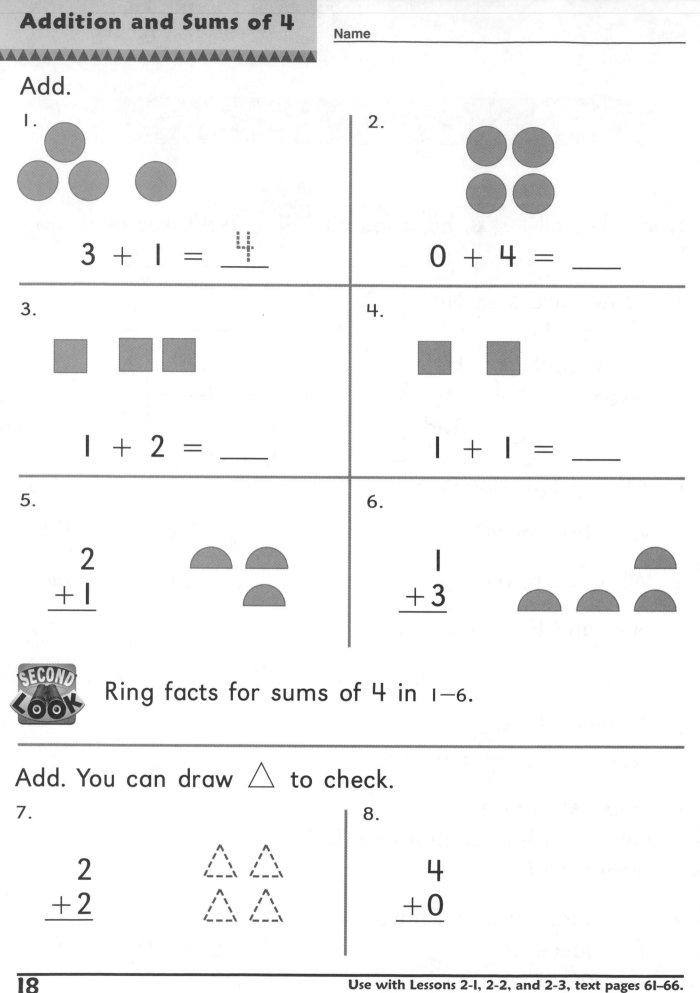

$3 + 1 = \underline{4}$

2.

$0 + 4 = \underline{}$

3.

$1 + 2 = \underline{}$

4.

$1 + 1 = \underline{}$

5.

$\begin{array}{r} 2 \\ +1 \\ \hline \end{array}$

6.

$\begin{array}{r} 1 \\ +3 \\ \hline \end{array}$

SECOND LOOK Ring facts for sums of 4 in 1–6.

Add. You can draw △ to check.

7.

$\begin{array}{r} 2 \\ +2 \\ \hline \end{array}$

8.

$\begin{array}{r} 4 \\ +0 \\ \hline \end{array}$

Use with Lessons 2-1, 2-2, and 2-3, text pages 61–66.

Find the sum.

1.
$$\begin{array}{r} 3 \\ +2 \\ \hline 5 \end{array}$$
$$\begin{array}{r} 0 \\ +2 \\ \hline \end{array}$$
$$\begin{array}{r} 1 \\ +3 \\ \hline \end{array}$$
$$\begin{array}{r} 6 \\ +0 \\ \hline \end{array}$$
$$\begin{array}{r} 1 \\ +1 \\ \hline \end{array}$$
$$\begin{array}{r} 1 \\ +2 \\ \hline \end{array}$$

2.
$$\begin{array}{r} 5 \\ +0 \\ \hline \end{array}$$
$$\begin{array}{r} 3 \\ +3 \\ \hline \end{array}$$
$$\begin{array}{r} 1 \\ +4 \\ \hline \end{array}$$
$$\begin{array}{r} 5 \\ +1 \\ \hline \end{array}$$
$$\begin{array}{r} 1 \\ +0 \\ \hline \end{array}$$
$$\begin{array}{r} 0 \\ +3 \\ \hline \end{array}$$

3.
$$\begin{array}{r} 2 \\ +3 \\ \hline \end{array}$$
$$\begin{array}{r} 4 \\ +0 \\ \hline \end{array}$$
$$\begin{array}{r} 0 \\ +1 \\ \hline \end{array}$$
$$\begin{array}{r} 3 \\ +1 \\ \hline \end{array}$$
$$\begin{array}{r} 0 \\ +5 \\ \hline \end{array}$$
$$\begin{array}{r} 0 \\ +4 \\ \hline \end{array}$$

4.
$$\begin{array}{r} 4 \\ +1 \\ \hline \end{array}$$
$$\begin{array}{r} 1 \\ +5 \\ \hline \end{array}$$
$$\begin{array}{r} 4 \\ +2 \\ \hline \end{array}$$
$$\begin{array}{r} 2 \\ +4 \\ \hline \end{array}$$
$$\begin{array}{r} 0 \\ +6 \\ \hline \end{array}$$
$$\begin{array}{r} 2 \\ +2 \\ \hline \end{array}$$

Add.

5. $2 + 2 = \underline{4}$ $0 + 6 = \underline{\hphantom{0}}$ $1 + 5 = \underline{\hphantom{0}}$

6. $2 + 0 = \underline{\hphantom{0}}$ $3 + 2 = \underline{\hphantom{0}}$ $2 + 1 = \underline{\hphantom{0}}$

7. $2 + 4 = \underline{\hphantom{0}}$ $4 + 1 = \underline{\hphantom{0}}$ $3 + 0 = \underline{\hphantom{0}}$

Name _____

Add. Change the order. Draw ○ to check.

1. 3 ●●●   0
   +0   +3 ○○○
   ___   ___
    3   3

2. 5
   +1   +___
   ___

3. 0
   +6   +___
   ___

4. 5
   +0   +___
   ___

5. 3 ●●●
   +1 ●   +___
   ___

6. 2 ●●
   +3 ●●●   +___
   ___

7. 0
   +4   +___
   ___

8. 1 ●
   +2 ●●   +___
   ___

Add.

9. 1 + 4 = ___   10. 4 + 2 = ___

11. 2 + 3 = ___   12. 2 + 2 = ___

13. 4 + 1 = ___   14. 2 + 4 = ___

SECOND LOOK In 9–14 ✔ related facts the same color.

Name _____

Look for patterns. Write the next fact.

1.
$$\begin{array}{r} 0 \\ +2 \\ \hline \end{array} \qquad \begin{array}{r} 1 \\ +2 \\ \hline \end{array} \qquad \begin{array}{r} 2 \\ +2 \\ \hline \end{array} \qquad \begin{array}{r} 3 \\ + \\ \hline \end{array} \qquad \begin{array}{r} \\ + \\ \hline \end{array}$$

2.
$$\begin{array}{r} 3 \\ +3 \\ \hline \end{array} \qquad \begin{array}{r} 2 \\ +3 \\ \hline \end{array} \qquad \begin{array}{r} 1 \\ +3 \\ \hline \end{array} \qquad \begin{array}{r} \\ + \\ \hline \end{array}$$

3.
$$\begin{array}{r} 5 \\ +1 \\ \hline \end{array} \quad \begin{array}{r} 4 \\ +1 \\ \hline \end{array} \quad \begin{array}{r} 3 \\ +1 \\ \hline \end{array} \quad \begin{array}{r} 2 \\ + \\ \hline \end{array} \quad \begin{array}{r} 1 \\ + \\ \hline \end{array} \quad \begin{array}{r} \\ + \\ \hline \end{array}$$

4.
$$\begin{array}{r} 1 \\ +0 \\ \hline \end{array} \quad \begin{array}{r} 2 \\ +0 \\ \hline \end{array} \quad \begin{array}{r} 3 \\ + \\ \hline \end{array} \quad \begin{array}{r} 4 \\ + \\ \hline \end{array} \quad \begin{array}{r} 5 \\ + \\ \hline \end{array} \quad \begin{array}{r} \\ + \\ \hline \end{array}$$

Add across. Add down.

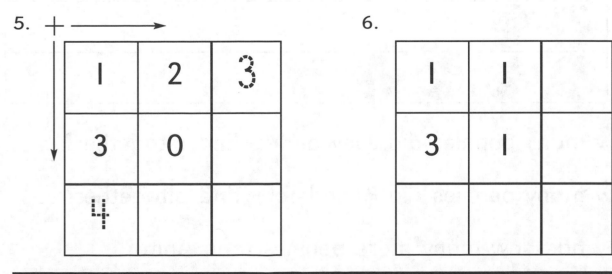

5.

+ →		
1	2	3
3	0	
4		

6.

1	1	
3	1	

Use with Lesson 2-7, text page 74.

1. Make a tally. Use the picture.

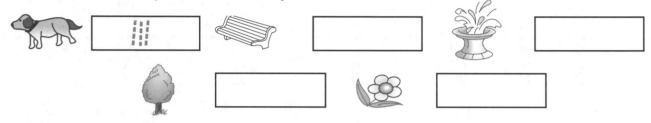

Use the picture graph for 2–4

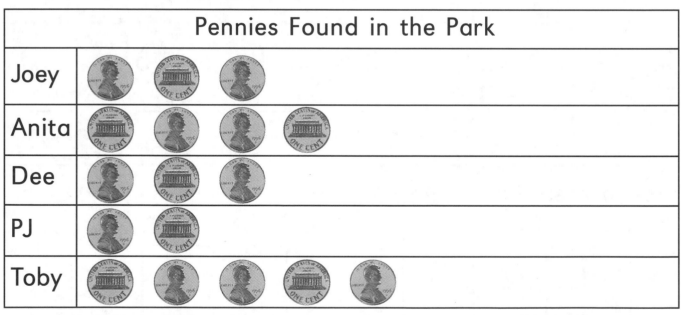

Pennies Found in the Park				
Joey				
Anita				
Dee				
PJ				
Toby				

2. How many pennies did Joey and PJ find altogether? _____

3. How many pennies did PJ and Toby find altogether? _____

4. Toby has how many more pennies than Anita? _____

Use with Lesson 2-9, text pages 77–78.

1. There are 4 🐱 playing. Then 2 more come to play. How many 🐱 play altogether?

_4_ ⊕ _2_ = ____

____ 🐱 altogether

2. Mark has 3 🐟. Then he gets 1 more 🐟. How many 🐟 does he have in all?

____ ◯ ____ = ____

____ 🐟 in all

3. Jason is third in line. There are 2 🚶 behind him. How many 🚶 are in line in all?

____ 🚶 in all

4. What flower comes next? Draw it.

Child	Number of Fish			
Sam	▨	▨		
Laura	▨			
Li	▨	▨		
Jackie	▨	▨	▨	▨

0   1   2   3   4

5. How many fish does the child with the most fish have?

____ fish

6. Sam and Laura have how many fish in all?

____ ◯ ____ = ____

____ fish in all.

Subtract.

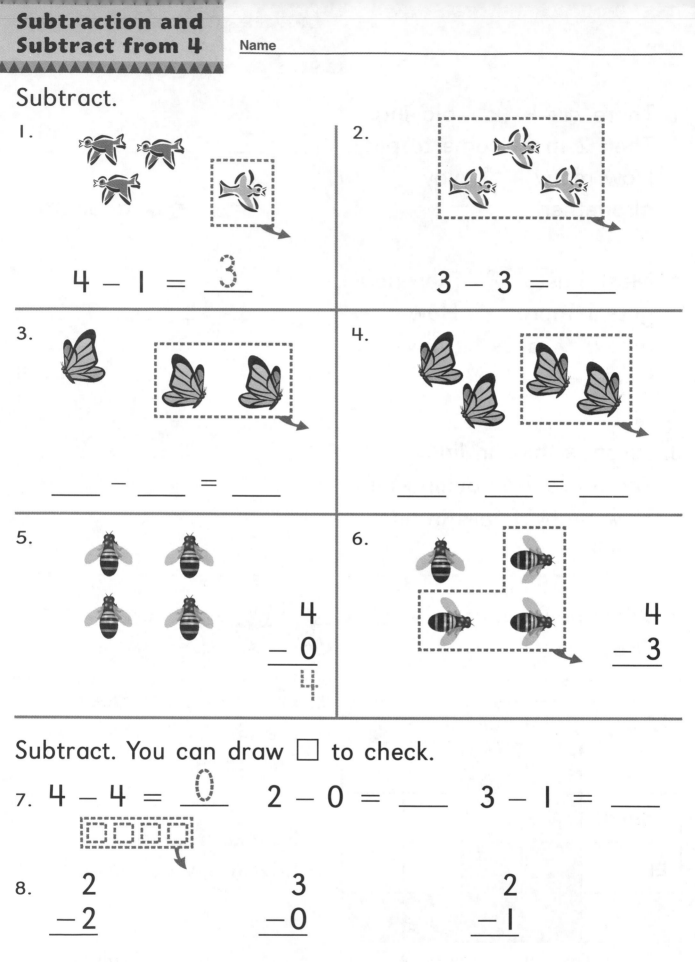

1. $4 - 1 = \underline{3}$

2. $3 - 3 = \underline{}$

3. $\underline{} - \underline{} = \underline{}$

4. $\underline{} - \underline{} = \underline{}$

5.
$$\begin{array}{r} 4 \\ -\ 0 \\ \hline 4 \end{array}$$

6.
$$\begin{array}{r} 4 \\ -\ 3 \\ \hline \end{array}$$

Subtract. You can draw ☐ to check.

7. $4 - 4 = \underline{0}$ $2 - 0 = \underline{}$ $3 - 1 = \underline{}$

8.
$$\begin{array}{r} 2 \\ -\ 2 \\ \hline \end{array} \qquad \begin{array}{r} 3 \\ -\ 0 \\ \hline \end{array} \qquad \begin{array}{r} 2 \\ -\ 1 \\ \hline \end{array}$$

Use with Lessons 3-1, 3-2, and 3-3, text pages 89–94.

Name _____

Find the difference.
Then color. ⟶

0–brown	1–yellow	2–red
3–blue	4–orange	5–green
6–purple		

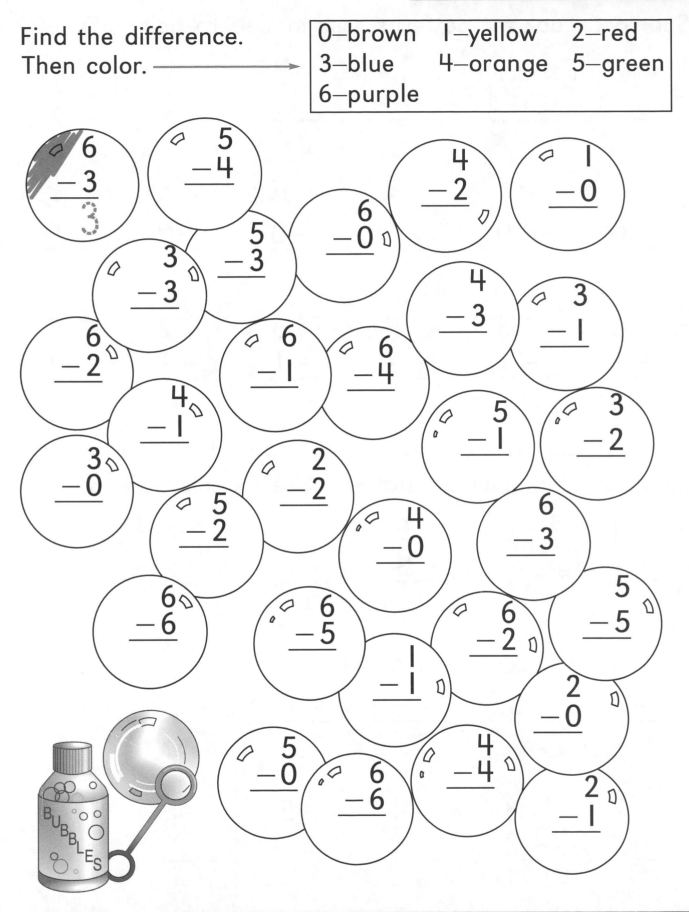

Use with Lessons 3-4 and 3-5, text pages 95–98.

Subtract. Look for patterns. Use models to help.

1.
$$\begin{array}{r} 6 \\ -3 \\ \hline 3 \end{array} \quad \square\square\square \atop \square\square\square \qquad \begin{array}{r} 5 \\ -3 \\ \hline \end{array} \qquad \begin{array}{r} 4 \\ -3 \\ \hline \end{array} \qquad \begin{array}{r} 3 \\ -3 \\ \hline \end{array}$$

2.
$$\begin{array}{r} 1 \\ -0 \\ \hline \end{array} \qquad \begin{array}{r} 2 \\ -0 \\ \hline \end{array} \qquad \begin{array}{r} 3 \\ -0 \\ \hline \end{array} \qquad \begin{array}{r} 4 \\ -0 \\ \hline \end{array} \qquad \begin{array}{r} 5 \\ -0 \\ \hline \end{array} \qquad \begin{array}{r} 6 \\ -0 \\ \hline \end{array}$$

3.
$$\begin{array}{r} 6 \\ -1 \\ \hline \end{array} \qquad \begin{array}{r} 5 \\ -1 \\ \hline \end{array} \qquad \begin{array}{r} 4 \\ -1 \\ \hline \end{array} \qquad \begin{array}{r} 3 \\ -1 \\ \hline \end{array} \qquad \begin{array}{r} 2 \\ -1 \\ \hline \end{array} \qquad \begin{array}{r} 1 \\ -1 \\ \hline \end{array}$$

Subtract. Complete the pattern. Use models to help.

4.
$$\begin{array}{r} 4 \\ -4 \\ \hline \end{array} \quad \square\square \atop \square\square \qquad \begin{array}{r} 5 \\ -4 \\ \hline \end{array} \qquad \begin{array}{r} \\ - \\ \hline \end{array}$$

5.
$$\begin{array}{r} 3 \\ -3 \\ \hline \end{array} \qquad \begin{array}{r} 4 \\ -3 \\ \hline \end{array} \qquad \begin{array}{r} 5 \\ - \\ \hline \end{array} \qquad \begin{array}{r} \\ - \\ \hline \end{array}$$

6.
$$\begin{array}{r} 2 \\ -2 \\ \hline \end{array} \qquad \begin{array}{r} 3 \\ -2 \\ \hline \end{array} \qquad \begin{array}{r} 4 \\ - \\ \hline \end{array} \qquad \begin{array}{r} 5 \\ - \\ \hline \end{array} \qquad \begin{array}{r} \\ - \\ \hline \end{array}$$

Use with Lesson 3-6, text page 100.

Write the related number sentences.

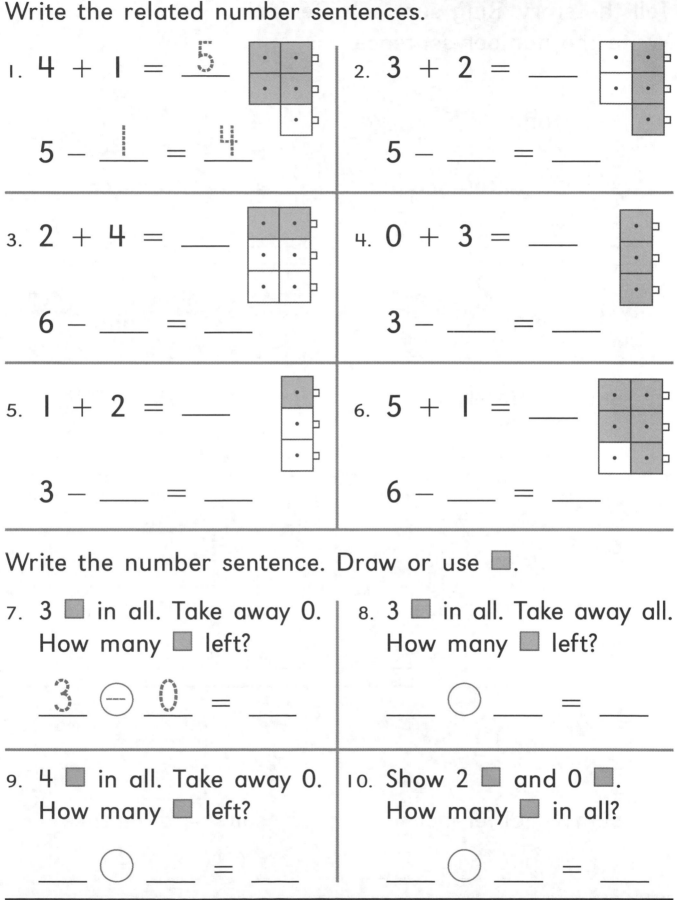

1. $4 + 1 = \underline{5}$

   $5 - \underline{1} = \underline{4}$

2. $3 + 2 = \underline{}$

   $5 - \underline{} = \underline{}$

3. $2 + 4 = \underline{}$

   $6 - \underline{} = \underline{}$

4. $0 + 3 = \underline{}$

   $3 - \underline{} = \underline{}$

5. $1 + 2 = \underline{}$

   $3 - \underline{} = \underline{}$

6. $5 + 1 = \underline{}$

   $6 - \underline{} = \underline{}$

Write the number sentence. Draw or use ■.

7. 3 ■ in all. Take away 0.
   How many ■ left?

   $\underline{3} \ominus \underline{0} = \underline{}$

8. 3 ■ in all. Take away all.
   How many ■ left?

   $\underline{} \bigcirc \underline{} = \underline{}$

9. 4 ■ in all. Take away 0.
   How many ■ left?

   $\underline{} \bigcirc \underline{} = \underline{}$

10. Show 2 ■ and 0 ■.
    How many ■ in all?

    $\underline{} \bigcirc \underline{} = \underline{}$

Use with Lessons 3-7 and 3-8, text pages 101–102.

27

Tell the story. Ring sum or difference.
Write the number sentence.

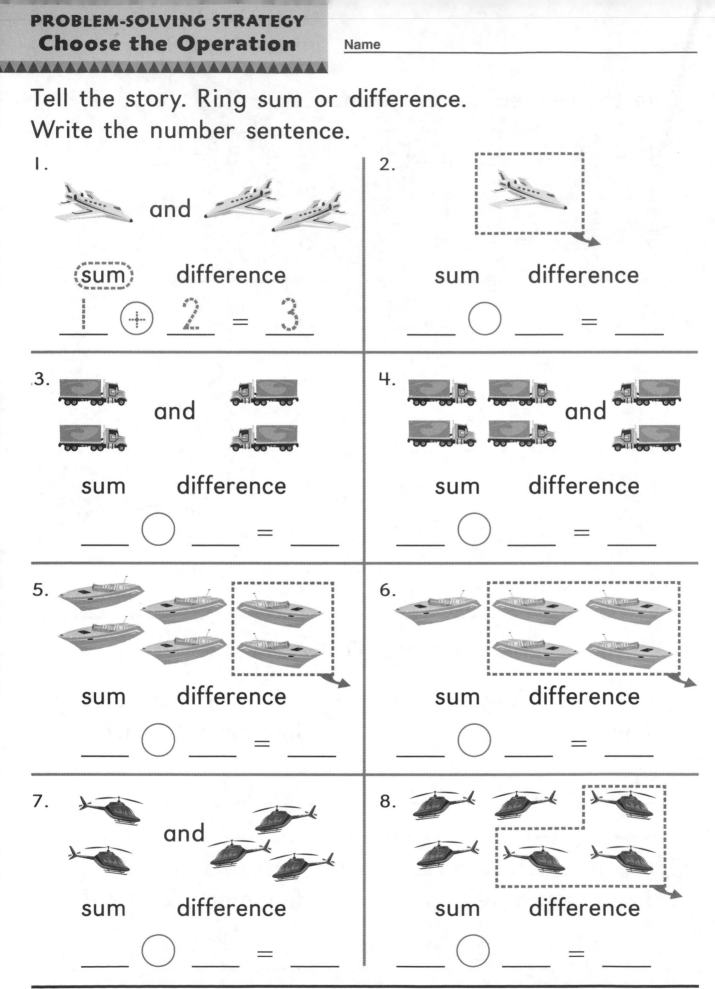

1. ✈ and ✈ ✈

   (sum)   difference

   1 (+) 2 = 3

2.

   sum   difference

   ___ ◯ ___ = ___

3. 🚚 and 🚚
   🚚   🚚

   sum   difference

   ___ ◯ ___ = ___

4. 🚚🚚 and 🚚
   🚚🚚   🚚

   sum   difference

   ___ ◯ ___ = ___

5.

   sum   difference

   ___ ◯ ___ = ___

6.

   sum   difference

   ___ ◯ ___ = ___

7. and

   sum   difference

   ___ ◯ ___ = ___

8.

   sum   difference

   ___ ◯ ___ = ___

Name _____

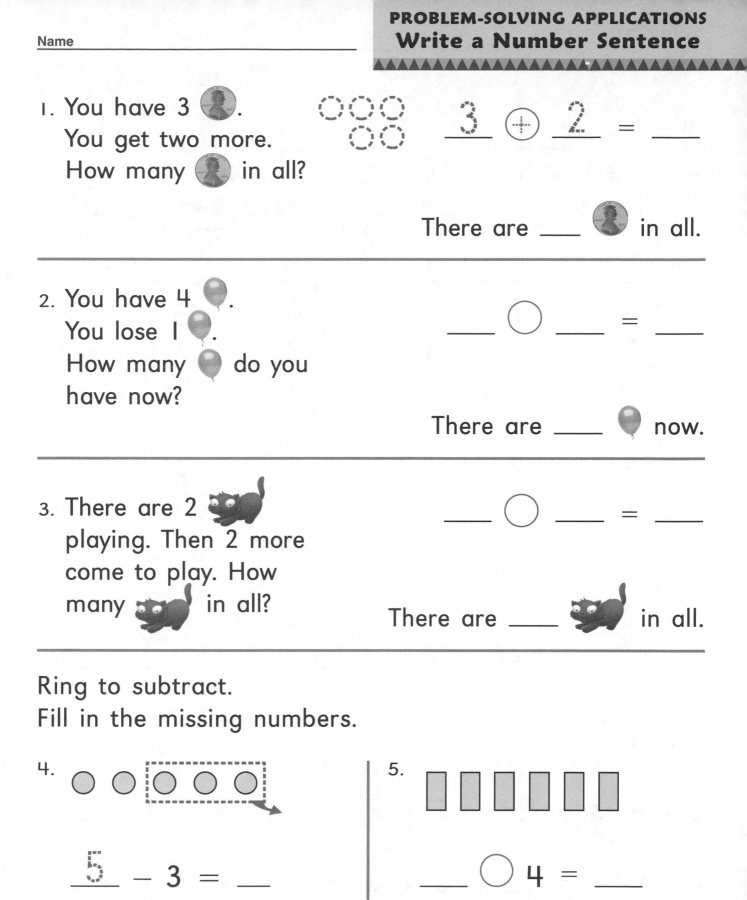

1. You have 3 🪙.
You get two more.
How many 🪙 in all?

__3__ ⊕ __2__ = ____

There are ____ 🪙 in all.

2. You have 4 🎈.
You lose 1 🎈.
How many 🎈 do you
have now?

____ ◯ ____ = ____

There are ____ 🎈 now.

3. There are 2 🐱
playing. Then 2 more
come to play. How
many 🐱 in all?

____ ◯ ____ = ____

There are ____ 🐱 in all.

Ring to subtract.
Fill in the missing numbers.

4.

__5__ – 3 = ____

____ ◯ are left.

5.

____ ◯ 4 = ____

____ ▯ are left.

Name _____

Add.

1.

$\begin{array}{r} 4 \\ +3 \\ \hline 7 \end{array}$

$\begin{array}{r} 3 \\ +4 \\ \hline \end{array}$

2.
$\begin{array}{r} 3 \\ +3 \\ \hline \end{array}$
$\begin{array}{r} 1 \\ +6 \\ \hline \end{array}$
$\begin{array}{r} 0 \\ +5 \\ \hline \end{array}$
$\begin{array}{r} 4 \\ +3 \\ \hline \end{array}$
$\begin{array}{r} 5 \\ +1 \\ \hline \end{array}$
$\begin{array}{r} 7 \\ +0 \\ \hline \end{array}$

3.
$\begin{array}{r} 3 \\ +1 \\ \hline \end{array}$
$\begin{array}{r} 2 \\ +3 \\ \hline \end{array}$
$\begin{array}{r} 3 \\ +4 \\ \hline \end{array}$
$\begin{array}{r} 6 \\ +0 \\ \hline \end{array}$
$\begin{array}{r} 1 \\ +4 \\ \hline \end{array}$
$\begin{array}{r} 4 \\ +2 \\ \hline \end{array}$

4.
$\begin{array}{r} 0 \\ +7 \\ \hline \end{array}$
$\begin{array}{r} 2 \\ +2 \\ \hline \end{array}$
$\begin{array}{r} 1 \\ +5 \\ \hline \end{array}$
$\begin{array}{r} 6 \\ +1 \\ \hline \end{array}$
$\begin{array}{r} 2 \\ +4 \\ \hline \end{array}$
$\begin{array}{r} 2 \\ +5 \\ \hline \end{array}$

Find the sum.

5. $1 + 3 = \underline{4}$ $5 + 2 = \underline{}$ $4 + 1 = \underline{}$

6. $3 + 4 = \underline{}$ $0 + 6 = \underline{}$ $6 + 1 = \underline{}$

7. $3 + 2 = \underline{}$ $4 + 3 = \underline{}$ $7 + 0 = \underline{}$

Use with Lesson 4-1, text pages 117–118.

Name _____

Find the sum.

1. ▪▪▪▪▪
    ▪
    ▪▪

    $$\begin{array}{r} 6 \\ +2 \\ \hline 8 \end{array}$$

    ▪▪
    ▪▪▪▪
    ▪

    $$\begin{array}{r} 2 \\ +6 \\ \hline \end{array}$$

2. $$\begin{array}{r} 1 \\ +6 \\ \hline \end{array}$$ $$\begin{array}{r} 4 \\ +4 \\ \hline \end{array}$$ $$\begin{array}{r} 2 \\ +4 \\ \hline \end{array}$$ $$\begin{array}{r} 7 \\ +1 \\ \hline \end{array}$$ $$\begin{array}{r} 3 \\ +2 \\ \hline \end{array}$$ $$\begin{array}{r} 0 \\ +8 \\ \hline \end{array}$$

3. $$\begin{array}{r} 6 \\ +2 \\ \hline \end{array}$$ $$\begin{array}{r} 2 \\ +5 \\ \hline \end{array}$$ $$\begin{array}{r} 3 \\ +5 \\ \hline \end{array}$$ $$\begin{array}{r} 5 \\ +0 \\ \hline \end{array}$$ $$\begin{array}{r} 5 \\ +3 \\ \hline \end{array}$$ $$\begin{array}{r} 3 \\ +4 \\ \hline \end{array}$$

4. $$\begin{array}{r} 8 \\ +0 \\ \hline \end{array}$$ $$\begin{array}{r} 1 \\ +4 \\ \hline \end{array}$$ $$\begin{array}{r} 5 \\ +2 \\ \hline \end{array}$$ $$\begin{array}{r} 2 \\ +6 \\ \hline \end{array}$$ $$\begin{array}{r} 4 \\ +2 \\ \hline \end{array}$$ $$\begin{array}{r} 2 \\ +3 \\ \hline \end{array}$$

Add.

5. $3 + 3 = \underline{6}$ $3 + 5 = \underline{}$ $6 + 1 = \underline{}$

6. $2 + 6 = \underline{}$ $0 + 7 = \underline{}$ $7 + 1 = \underline{}$

7. $5 + 1 = \underline{}$ $6 + 2 = \underline{}$ $4 + 4 = \underline{}$

Name _____

Find the sum.

1.
$$\begin{array}{r} 6 \\ +3 \\ \hline 9 \end{array}$$

$$\begin{array}{r} 3 \\ +6 \\ \hline \end{array}$$

2.
$$\begin{array}{r} 6 \\ +0 \\ \hline 6 \end{array}$$
$$\begin{array}{r} 1 \\ +8 \\ \hline \end{array}$$
$$\begin{array}{r} 5 \\ +3 \\ \hline \end{array}$$
$$\begin{array}{r} 2 \\ +7 \\ \hline \end{array}$$
$$\begin{array}{r} 3 \\ +3 \\ \hline \end{array}$$
$$\begin{array}{r} 2 \\ +5 \\ \hline \end{array}$$

3.
$$\begin{array}{r} 6 \\ +3 \\ \hline \end{array}$$
$$\begin{array}{r} 7 \\ +0 \\ \hline \end{array}$$
$$\begin{array}{r} 2 \\ +6 \\ \hline \end{array}$$
$$\begin{array}{r} 5 \\ +1 \\ \hline \end{array}$$
$$\begin{array}{r} 0 \\ +9 \\ \hline \end{array}$$
$$\begin{array}{r} 4 \\ +4 \\ \hline \end{array}$$

4.
$$\begin{array}{r} 1 \\ +6 \\ \hline \end{array}$$
$$\begin{array}{r} 7 \\ +2 \\ \hline \end{array}$$
$$\begin{array}{r} 2 \\ +4 \\ \hline \end{array}$$
$$\begin{array}{r} 5 \\ +2 \\ \hline \end{array}$$
$$\begin{array}{r} 8 \\ +1 \\ \hline \end{array}$$
$$\begin{array}{r} 3 \\ +5 \\ \hline \end{array}$$

Add.

5. $4 + 2 = \underline{6}$ $3 + 4 = \underline{\hspace{1.5em}}$ $3 + 6 = \underline{\hspace{1.5em}}$

6. $5 + 4 = \underline{\hspace{1.5em}}$ $7 + 1 = \underline{\hspace{1.5em}}$ $2 + 7 = \underline{\hspace{1.5em}}$

7. $6 + 2 = \underline{\hspace{1.5em}}$ $0 + 6 = \underline{\hspace{1.5em}}$ $6 + 1 = \underline{\hspace{1.5em}}$

Use with Lesson 4-3, text pages 121–122.

Name _____

Add.

1.
$$\begin{array}{r} 2 \\ +8 \\ \hline 10 \end{array}$$
$$\begin{array}{r} 8 \\ +2 \\ \hline \end{array}$$

2.
$$\begin{array}{r} 6 \\ +3 \\ \hline 9 \end{array}$$
$$\begin{array}{r} 4 \\ +6 \\ \hline \end{array}$$
$$\begin{array}{r} 3 \\ +5 \\ \hline \end{array}$$
$$\begin{array}{r} 1 \\ +6 \\ \hline \end{array}$$
$$\begin{array}{r} 8 \\ +2 \\ \hline \end{array}$$
$$\begin{array}{r} 4 \\ +5 \\ \hline \end{array}$$

3.
$$\begin{array}{r} 7 \\ +3 \\ \hline \end{array}$$
$$\begin{array}{r} 1 \\ +8 \\ \hline \end{array}$$
$$\begin{array}{r} 5 \\ +2 \\ \hline \end{array}$$
$$\begin{array}{r} 2 \\ +6 \\ \hline \end{array}$$
$$\begin{array}{r} 9 \\ +0 \\ \hline \end{array}$$
$$\begin{array}{r} 1 \\ +9 \\ \hline \end{array}$$

4.
$$\begin{array}{r} 5 \\ +4 \\ \hline \end{array}$$
$$\begin{array}{r} 2 \\ +7 \\ \hline \end{array}$$
$$\begin{array}{r} 5 \\ +3 \\ \hline \end{array}$$
$$\begin{array}{r} 5 \\ +5 \\ \hline \end{array}$$
$$\begin{array}{r} 3 \\ +6 \\ \hline \end{array}$$
$$\begin{array}{r} 0 \\ +8 \\ \hline \end{array}$$

Find the sum.

5. $3 + 4 = 7$ $8 + 2 = \underline{}$ $3 + 7 = \underline{}$

6. $4 + 6 = \underline{}$ $7 + 3 = \underline{}$ $6 + 3 = \underline{}$

7. $9 + 1 = \underline{}$ $6 + 2 = \underline{}$ $4 + 4 = \underline{}$

SECOND LOOK In 5–7 ring any related facts.

Add.

1.

$$\begin{array}{r} 6 \\ +5 \\ \hline \end{array}$$

$$\begin{array}{r} 5 \\ +6 \\ \hline \end{array}$$

2.

$$\begin{array}{r} 7 \\ +4 \\ \hline \end{array}$$
$$\begin{array}{r} 5 \\ +6 \\ \hline \end{array}$$
$$\begin{array}{r} 2 \\ +5 \\ \hline \end{array}$$
$$\begin{array}{r} 2 \\ +9 \\ \hline \end{array}$$
$$\begin{array}{r} 1 \\ +7 \\ \hline \end{array}$$
$$\begin{array}{r} 5 \\ +5 \\ \hline \end{array}$$

3.

$$\begin{array}{r} 9 \\ +0 \\ \hline \end{array}$$
$$\begin{array}{r} 8 \\ +2 \\ \hline \end{array}$$
$$\begin{array}{r} 6 \\ +5 \\ \hline \end{array}$$
$$\begin{array}{r} 1 \\ +6 \\ \hline \end{array}$$
$$\begin{array}{r} 9 \\ +2 \\ \hline \end{array}$$
$$\begin{array}{r} 3 \\ +6 \\ \hline \end{array}$$

4.

$$\begin{array}{r} 3 \\ +8 \\ \hline \end{array}$$
$$\begin{array}{r} 6 \\ +3 \\ \hline \end{array}$$
$$\begin{array}{r} 2 \\ +7 \\ \hline \end{array}$$
$$\begin{array}{r} 3 \\ +7 \\ \hline \end{array}$$
$$\begin{array}{r} 2 \\ +6 \\ \hline \end{array}$$
$$\begin{array}{r} 2 \\ +8 \\ \hline \end{array}$$

Find the sum.

5. $7 + 3 = \underline{10}$ $3 + 5 = \underline{}$ $1 + 9 = \underline{}$

6. $4 + 7 = \underline{}$ $7 + 2 = \underline{}$ $8 + 3 = \underline{}$

7. $8 + 1 = \underline{}$ $4 + 4 = \underline{}$ $0 + 7 = \underline{}$

Use with Lesson 4-5, text pages 125–126.

Find the sum.

1.
$$\begin{array}{r} 7 \\ +5 \\ \hline 12 \end{array}$$

$$\begin{array}{r} 5 \\ +7 \\ \hline \end{array}$$

2.
$$\begin{array}{r} 6 \\ +6 \\ \hline 12 \end{array}$$
$$\begin{array}{r} 0 \\ +9 \\ \hline \end{array}$$
$$\begin{array}{r} 5 \\ +5 \\ \hline \end{array}$$
$$\begin{array}{r} 6 \\ +5 \\ \hline \end{array}$$
$$\begin{array}{r} 5 \\ +7 \\ \hline \end{array}$$
$$\begin{array}{r} 2 \\ +7 \\ \hline \end{array}$$

3.
$$\begin{array}{r} 9 \\ +1 \\ \hline \end{array}$$
$$\begin{array}{r} 4 \\ +8 \\ \hline \end{array}$$
$$\begin{array}{r} 5 \\ +4 \\ \hline \end{array}$$
$$\begin{array}{r} 9 \\ +3 \\ \hline \end{array}$$
$$\begin{array}{r} 3 \\ +8 \\ \hline \end{array}$$
$$\begin{array}{r} 6 \\ +4 \\ \hline \end{array}$$

4.
$$\begin{array}{r} 1 \\ +8 \\ \hline \end{array}$$
$$\begin{array}{r} 4 \\ +7 \\ \hline \end{array}$$
$$\begin{array}{r} 3 \\ +9 \\ \hline \end{array}$$
$$\begin{array}{r} 5 \\ +3 \\ \hline \end{array}$$
$$\begin{array}{r} 4 \\ +6 \\ \hline \end{array}$$
$$\begin{array}{r} 2 \\ +9 \\ \hline \end{array}$$

Find the sum.

5. $8 + 4 = 12$ $6 + 3 = \underline{\quad}$ $7 + 4 = \underline{\quad}$

6. $6 + 2 = \underline{\quad}$ $7 + 5 = \underline{\quad}$ $6 + 6 = \underline{\quad}$

7. $8 + 3 = \underline{\quad}$ $5 + 6 = \underline{\quad}$ $8 + 2 = \underline{\quad}$

SECOND LOOK In 5–7 ring sums greater than 10.

Write the addition sentence for each 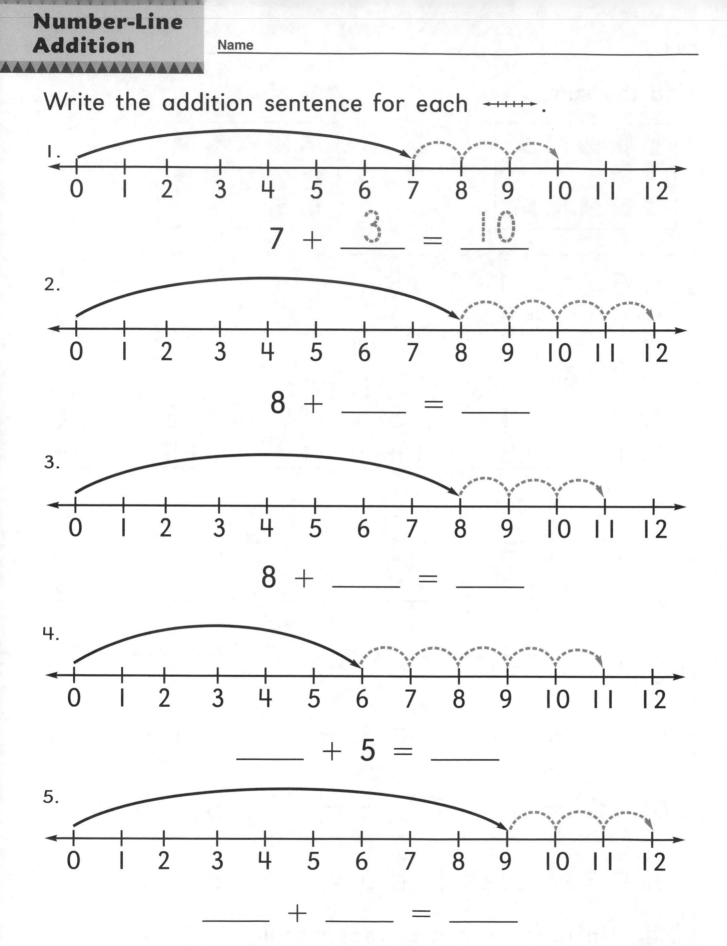.

1.

7 + __3__ = __10__

2.

8 + ____ = ____

3.

8 + ____ = ____

4.

____ + 5 = ____

5.

____ + ____ = ____

Find the sum. Complete the pattern.

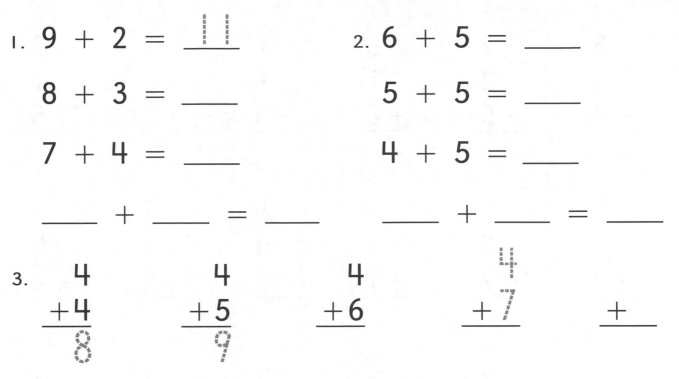

1. $9 + 2 = 11$

$8 + 3 = \underline{\quad}$

$7 + 4 = \underline{\quad}$

$\underline{\quad} + \underline{\quad} = \underline{\quad}$

2. $6 + 5 = \underline{\quad}$

$5 + 5 = \underline{\quad}$

$4 + 5 = \underline{\quad}$

$\underline{\quad} + \underline{\quad} = \underline{\quad}$

3.
$$\begin{array}{r} 4 \\ +4 \\ \hline 8 \end{array} \qquad \begin{array}{r} 4 \\ +5 \\ \hline 9 \end{array} \qquad \begin{array}{r} 4 \\ +6 \\ \hline \end{array} \qquad \begin{array}{r} 4 \\ +7 \\ \hline \end{array} \qquad \begin{array}{r} \\ + \\ \hline \end{array}$$

Draw the part joined in each doubles fact.
Complete the addition sentence.

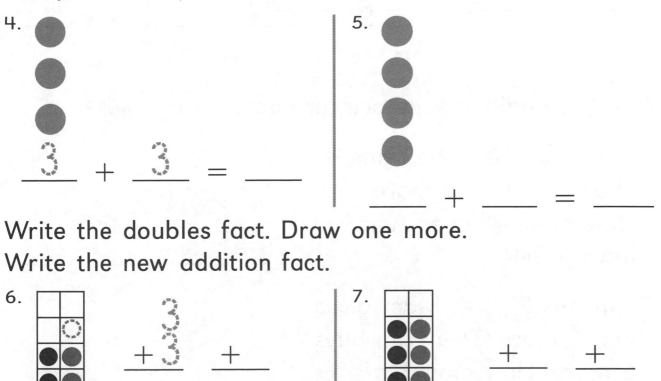

4. $\underline{3} + \underline{3} = \underline{\quad}$

5. $\underline{\quad} + \underline{\quad} = \underline{\quad}$

Write the doubles fact. Draw one more.
Write the new addition fact.

6. $\begin{array}{r} 3 \\ + \\ \hline \end{array} + \underline{\quad}$

7. $\underline{\quad} + \underline{\quad}$

Use with Lessons 4-8, 4-9, text pages 131–134.

Add.

1.
$$\begin{array}{r} 5 \\ 3 \\ +3 \\ \hline \end{array}$$
$$\begin{array}{r} 2 \\ 2 \\ +5 \\ \hline \end{array}$$
$$\begin{array}{r} 3 \\ 1 \\ +8 \\ \hline \end{array}$$
$$\begin{array}{r} 1 \\ 1 \\ +9 \\ \hline \end{array}$$
$$\begin{array}{r} 1 \\ 6 \\ +3 \\ \hline \end{array}$$
$$\begin{array}{r} 2 \\ 1 \\ +5 \\ \hline \end{array}$$

2.
$$\begin{array}{r} 2 \\ 0 \\ +7 \\ \hline \end{array}$$
$$\begin{array}{r} 3 \\ 2 \\ +7 \\ \hline \end{array}$$
$$\begin{array}{r} 2 \\ 1 \\ +7 \\ \hline \end{array}$$
$$\begin{array}{r} 5 \\ 4 \\ +3 \\ \hline \end{array}$$
$$\begin{array}{r} 3 \\ 1 \\ +4 \\ \hline \end{array}$$
$$\begin{array}{r} 3 \\ 4 \\ +4 \\ \hline \end{array}$$

3.
$$\begin{array}{r} 3 \\ 1 \\ +7 \\ \hline \end{array}$$
$$\begin{array}{r} 4 \\ 0 \\ +4 \\ \hline \end{array}$$
$$\begin{array}{r} 4 \\ 1 \\ +6 \\ \hline \end{array}$$
$$\begin{array}{r} 2 \\ 6 \\ +2 \\ \hline \end{array}$$
$$\begin{array}{r} 1 \\ 3 \\ +4 \\ \hline \end{array}$$
$$\begin{array}{r} 1 \\ 2 \\ +7 \\ \hline \end{array}$$

Write the addition sentence for each.

4. Ann has 3 ✿. She finds 4.
Then she finds 5 more.
How many ✿ does Ann
have in all? ____ + ____ + ____ = ____

5. Tim has 2 🚚. Mom gives
him 3 more. Then Tim buys
6 more. How many 🚚
does Tim have altogether? ____ + ____ + ____ = ____

Use with Lessons 4-10, 4-11, text pages 135–138.

Name _____

Add. Use a strategy you know.

| Count on. | Look for patterns. |
| Use doubles. | Think 10. |

1. $\begin{array}{r} 3 \\ +8 \\ \hline \end{array}$ $\begin{array}{r} 7 \\ +5 \\ \hline \end{array}$ $\begin{array}{r} 4 \\ +6 \\ \hline \end{array}$ $\begin{array}{r} 3 \\ +5 \\ \hline \end{array}$ $\begin{array}{r} 2 \\ +8 \\ \hline \end{array}$ $\begin{array}{r} 8¢ \\ +4¢ \\ \hline \end{array}$ ¢

2. $\begin{array}{r} 2 \\ +9 \\ \hline \end{array}$ $\begin{array}{r} 3 \\ +6 \\ \hline \end{array}$ $\begin{array}{r} 7 \\ +5 \\ \hline \end{array}$ $\begin{array}{r} 6 \\ +2 \\ \hline \end{array}$ $\begin{array}{r} 5 \\ +4 \\ \hline \end{array}$ $\begin{array}{r} 5¢ \\ +6¢ \\ \hline \end{array}$ ¢

3. $\begin{array}{r} 3 \\ +7 \\ \hline \end{array}$ $\begin{array}{r} 9 \\ +2 \\ \hline \end{array}$ $\begin{array}{r} 9 \\ +3 \\ \hline \end{array}$ $\begin{array}{r} 7 \\ +4 \\ \hline \end{array}$ $\begin{array}{r} 6 \\ +3 \\ \hline \end{array}$ $\begin{array}{r} 4¢ \\ +4¢ \\ \hline \end{array}$ ¢

4. $\begin{array}{r} 4 \\ +7 \\ \hline \end{array}$ $\begin{array}{r} 6 \\ +6 \\ \hline \end{array}$ $\begin{array}{r} 5 \\ +3 \\ \hline \end{array}$ $\begin{array}{r} 2 \\ +7 \\ \hline \end{array}$ $\begin{array}{r} 4¢ \\ +7¢ \\ \hline \end{array}$ ¢ $\begin{array}{r} 3¢ \\ +7¢ \\ \hline \end{array}$ ¢

5. $\begin{array}{r} 0 \\ +7 \\ \hline \end{array}$ $\begin{array}{r} 5 \\ +6 \\ \hline \end{array}$ $\begin{array}{r} 5 \\ +5 \\ \hline \end{array}$ $\begin{array}{r} 3 \\ +6 \\ \hline \end{array}$ $\begin{array}{r} 5¢ \\ +7¢ \\ \hline \end{array}$ ¢ $\begin{array}{r} 3¢ \\ +5¢ \\ \hline \end{array}$ ¢

6. $7 + 2 + 3 =$ _____ $3¢ + 4¢ + 3¢ =$ _____ ¢

7. $6 + 1 + 4 =$ _____ $2¢ + 3¢ + 6¢ =$ _____ ¢

Use with Lesson 4-12, text pages 139–140.

Choose the question.
Write the number sentence.

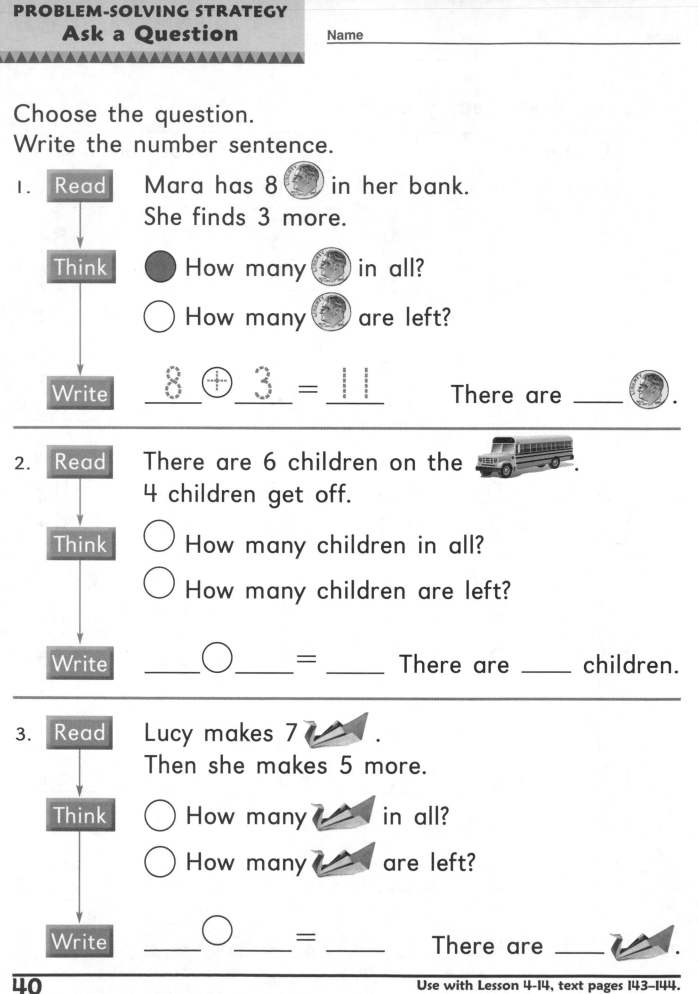

1. **Read** Mara has 8 🪙 in her bank.
She finds 3 more.

   Think ⬤ How many 🪙 in all?

   ◯ How many 🪙 are left?

   Write _8_ ⊕ _3_ = _11_   There are ____ 🪙.

2. **Read** There are 6 children on the 🚌.
4 children get off.

   Think ◯ How many children in all?

   ◯ How many children are left?

   Write ____◯____ = ____ There are ____ children.

3. **Read** Lucy makes 7 ✈.
Then she makes 5 more.

   Think ◯ How many ✈ in all?

   ◯ How many ✈ are left?

   Write ____◯____ = ____   There are ____ ✈.

40

Name _____

Read	Think	Ring	Write

1. Bill has 5 🦕. He got 7 more. How many 🦕 does he have in all?

(add) or subtract

⊕
−

$$\begin{array}{r} 5 \\ +7 \\ \hline 12 \end{array}$$

__12__ in all

2. There were 6 🧁. 3 were eaten. How many 🧁 are left?

add or subtract

+
−

____ are left

3. There are 6 🦆 on the lake. 5 more flew in. How many 🦆 in all?

add or subtract

+
−

____ in all

4. There are 8 🐌 in the yard. Sue finds 4 more. How many 🐌 now?

add or subtract

+
−

____ now

1. Write other names for 9.

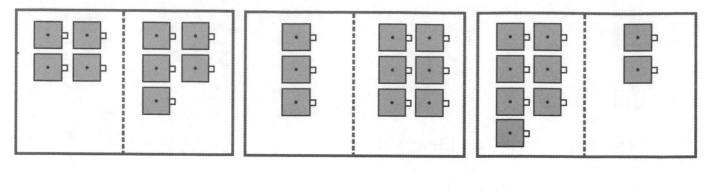

_____ + _____   _____ + _____   _____ + _____

2. Write other names for 11.

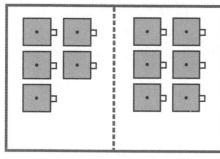 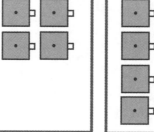 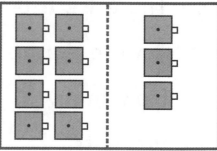

_____ + _____   _____ + _____   _____ + _____

3. Write other names for 1.

4	5	6
—	—	—
___	___	___

4. Write other names for 2.

4	5	6
—	—	—
___	___	___

5. Write other names for 3.

6	5	4
—	—	—
___	___	___

6. Write other names for 0.

6	4	2
—	—	—
___	___	___

Use with Lesson 5-1, text pages 155–156.

Name _____

Subtract.

1.
	7		7
	− 5		− 2
	2		

2.

5	7	6	7	4	6
− 4	− 3	− 4	− 0	− 2	− 1

3.

4	6	7	5	7	5
− 3	− 2	− 7	− 2	− 1	− 0

4.

7	6	5	7	4	7
− 6	− 6	− 3	− 5	− 0	− 4

Find the difference.

5. $6 − 5 =$ __1__ $7 − 4 =$ ___ $6 − 3 =$ ___

6. $7 − 2 =$ ___ $5 − 5 =$ ___ $7 − 3 =$ ___

7. $6 − 2 =$ ___ $7 − 0 =$ ___ $7 − 6 =$ ___

Use with Lesson 5-2, text pages 157–158.

Find the difference.

1.
$$
\begin{array}{r} 8 \\ -\ 3 \\ \hline 5 \end{array}
\qquad
\begin{array}{r} 8 \\ -\ 5 \\ \hline \end{array}
$$

2.
$$
\begin{array}{r} 7 \\ -\ 1 \\ \hline \end{array}
\quad
\begin{array}{r} 8 \\ -\ 4 \\ \hline \end{array}
\quad
\begin{array}{r} 5 \\ -\ 1 \\ \hline \end{array}
\quad
\begin{array}{r} 8 \\ -\ 2 \\ \hline \end{array}
\quad
\begin{array}{r} 6 \\ -\ 0 \\ \hline \end{array}
\quad
\begin{array}{r} 8 \\ -\ 7 \\ \hline \end{array}
$$

3.
$$
\begin{array}{r} 8 \\ -\ 8 \\ \hline \end{array}
\quad
\begin{array}{r} 6 \\ -\ 4 \\ \hline \end{array}
\quad
\begin{array}{r} 7 \\ -\ 5 \\ \hline \end{array}
\quad
\begin{array}{r} 8 \\ -\ 3 \\ \hline \end{array}
\quad
\begin{array}{r} 5 \\ -\ 3 \\ \hline \end{array}
\quad
\begin{array}{r} 8 \\ -\ 0 \\ \hline \end{array}
$$

4.
$$
\begin{array}{r} 7 \\ -\ 3 \\ \hline \end{array}
\quad
\begin{array}{r} 8 \\ -\ 6 \\ \hline \end{array}
\quad
\begin{array}{r} 5 \\ -\ 4 \\ \hline \end{array}
\quad
\begin{array}{r} 7 \\ -\ 0 \\ \hline \end{array}
\quad
\begin{array}{r} 8 \\ -\ 1 \\ \hline \end{array}
\quad
\begin{array}{r} 6 \\ -\ 3 \\ \hline \end{array}
$$

Subtract.

5. $7 - 7 = \underline{\ 0\ }$ $\qquad$ $8 - 2 = \underline{\ \ \ }$ $\qquad$ $8 - 8 = \underline{\ \ \ }$

6. $6 - 5 = \underline{\ \ \ }$ $\qquad$ $8 - 4 = \underline{\ \ \ }$ $\qquad$ $7 - 2 = \underline{\ \ \ }$

7. $8 - 5 = \underline{\ \ \ }$ $\qquad$ $7 - 4 = \underline{\ \ \ }$ $\qquad$ $8 - 7 = \underline{\ \ \ }$

Use with Lesson 5-3, text pages 159–160.

Name _____

Find the difference.

1.

$9 - 7 = \underline{2}$ $9 - 2 = \underline{}$

2. $9 - 1 = \underline{8}$ $9 - 4 = \underline{}$ $8 - 6 = \underline{}$

3. $7 - 7 = \underline{}$ $9 - 8 = \underline{}$ $9 - 6 = \underline{}$

4. $9 - 3 = \underline{}$ $7 - 5 = \underline{}$ $9 - 9 = \underline{}$

Subtract.

5. **7**

8	9	7	9	6	9
$-\ 1$	$-\ 5$	$-\ 6$	$-\ 2$	$-\ 6$	$-\ 7$
7					

6. **1**

8	9	8	7	6	9
$-\ 7$	$-\ 6$	$-\ 6$	$-\ 6$	$-\ 3$	$-\ 8$

7. **4**

9	7	8	9	8	7
$-\ 1$	$-\ 5$	$-\ 4$	$-\ 5$	$-\ 5$	$-\ 2$

SECOND LOOK In 5–7 ring facts that equal the number on each kite.

Use with Lesson 5-4, text pages 161–162.

Find the difference.

1.

$10 - 6 = \underline{4}$ $10 - 4 = \underline{}$

2. $10 - 8 = \underline{2}$ $10 - 3 = \underline{}$ $9 - 8 = \underline{}$

3. $8 - 6 = \underline{}$ $7 - 0 = \underline{}$ $10 - 6 = \underline{}$

4. $10 - 9 = \underline{}$ $9 - 2 = \underline{}$ $10 - 7 = \underline{}$

Subtract.

5.
$\begin{array}{r} 9 \\ -\ 4 \\ \hline 5 \end{array}$
$\begin{array}{r} 10 \\ -\ 5 \\ \hline \end{array}$
$\begin{array}{r} 8 \\ -\ 2 \\ \hline \end{array}$
$\begin{array}{r} 10 \\ -\ 4 \\ \hline \end{array}$
$\begin{array}{r} 7 \\ -\ 1 \\ \hline \end{array}$
$\begin{array}{r} 10 \\ -\ 8 \\ \hline \end{array}$

6.
$\begin{array}{r} 8 \\ -\ 3 \\ \hline \end{array}$
$\begin{array}{r} 9 \\ -\ 6 \\ \hline \end{array}$
$\begin{array}{r} 10 \\ -\ 1 \\ \hline \end{array}$
$\begin{array}{r} 7 \\ -\ 5 \\ \hline \end{array}$
$\begin{array}{r} 10 \\ -\ 3 \\ \hline \end{array}$
$\begin{array}{r} 9 \\ -\ 5 \\ \hline \end{array}$

7.
$\begin{array}{r} 10 \\ -\ 9 \\ \hline \end{array}$
$\begin{array}{r} 9 \\ -\ 7 \\ \hline \end{array}$
$\begin{array}{r} 7 \\ -\ 3 \\ \hline \end{array}$
$\begin{array}{r} 10 \\ -\ 2 \\ \hline \end{array}$
$\begin{array}{r} 8 \\ -\ 1 \\ \hline \end{array}$
$\begin{array}{r} 9 \\ -\ 9 \\ \hline \end{array}$

8.
$\begin{array}{r} 8 \\ -\ 8 \\ \hline \end{array}$
$\begin{array}{r} 10 \\ -\ 7 \\ \hline \end{array}$
$\begin{array}{r} 9 \\ -\ 3 \\ \hline \end{array}$
$\begin{array}{r} 7 \\ -\ 2 \\ \hline \end{array}$
$\begin{array}{r} 10 \\ -\ 5 \\ \hline \end{array}$
$\begin{array}{r} 8 \\ -\ 5 \\ \hline \end{array}$

Use with Lesson 5-5, text pages 163–164.

Name _____

Find the difference.

1.

$11 - 7 = \underline{4}$

$11 - 4 = \underline{}$

2.

$11 - 3 = \underline{8}$ $11 - 8 = \underline{3}$

3.

$11 - 9 = \underline{}$ $11 - 2 = \underline{}$

4.

$11 - 5 = \underline{}$ $11 - 6 = \underline{}$

5.

$10 - 6 = \underline{}$ $10 - 4 = \underline{}$

Subtract.

6.
$$\begin{array}{r} 11 \\ -\ 6 \\ \hline 5 \end{array}$$
$$\begin{array}{r} 10 \\ -\ 3 \\ \hline \end{array}$$
$$\begin{array}{r} 8 \\ -\ 4 \\ \hline \end{array}$$
$$\begin{array}{r} 11 \\ -\ 7 \\ \hline \end{array}$$
$$\begin{array}{r} 7 \\ -\ 0 \\ \hline \end{array}$$
$$\begin{array}{r} 11 \\ -\ 2 \\ \hline \end{array}$$

7.
$$\begin{array}{r} 9 \\ -\ 3 \\ \hline \end{array}$$
$$\begin{array}{r} 11 \\ -\ 8 \\ \hline \end{array}$$
$$\begin{array}{r} 10 \\ -\ 1 \\ \hline \end{array}$$
$$\begin{array}{r} 8 \\ -\ 3 \\ \hline \end{array}$$
$$\begin{array}{r} 11 \\ -\ 5 \\ \hline \end{array}$$
$$\begin{array}{r} 9 \\ -\ 7 \\ \hline \end{array}$$

8.
$$\begin{array}{r} 11 \\ -\ 4 \\ \hline \end{array}$$
$$\begin{array}{r} 9 \\ -\ 8 \\ \hline \end{array}$$
$$\begin{array}{r} 10 \\ -\ 7 \\ \hline \end{array}$$
$$\begin{array}{r} 11 \\ -\ 3 \\ \hline \end{array}$$
$$\begin{array}{r} 10 \\ -\ 5 \\ \hline \end{array}$$
$$\begin{array}{r} 7 \\ -\ 2 \\ \hline \end{array}$$

9.
$$\begin{array}{r} 10 \\ -\ 2 \\ \hline \end{array}$$
$$\begin{array}{r} 9 \\ -\ 6 \\ \hline \end{array}$$
$$\begin{array}{r} 9 \\ -\ 4 \\ \hline \end{array}$$
$$\begin{array}{r} 7 \\ -\ 3 \\ \hline \end{array}$$
$$\begin{array}{r} 10 \\ -\ 6 \\ \hline \end{array}$$
$$\begin{array}{r} 9 \\ -\ 5 \\ \hline \end{array}$$

Use with Lesson 5-6, text pages 165–166.

Name _____

Find the difference.

1.

$12 - 8 =$ __4__ $12 - 4 =$ ___

2.
$12 - 3 =$ __9__ $12 - 9 =$ __3__

3.
$12 - 5 =$ __ $12 - 7 =$ __

4.
$11 - 5 =$ __ $11 - 6 =$ __

5.
$10 - 3 =$ __ $10 - 7 =$ __

Subtract.

6.
$$\begin{array}{r} 12 \\ -\ 5 \\ \hline 7 \end{array}$$
$$\begin{array}{r} 10 \\ -\ 2 \\ \hline \end{array}$$
$$\begin{array}{r} 11 \\ -\ 7 \\ \hline \end{array}$$
$$\begin{array}{r} 9 \\ -\ 6 \\ \hline \end{array}$$
$$\begin{array}{r} 12 \\ -\ 8 \\ \hline \end{array}$$
$$\begin{array}{r} 11 \\ -\ 6 \\ \hline \end{array}$$

7.
$$\begin{array}{r} 10 \\ -\ 9 \\ \hline \end{array}$$
$$\begin{array}{r} 9 \\ -\ 5 \\ \hline \end{array}$$
$$\begin{array}{r} 12 \\ -\ 9 \\ \hline \end{array}$$
$$\begin{array}{r} 11 \\ -\ 8 \\ \hline \end{array}$$
$$\begin{array}{r} 10 \\ -\ 7 \\ \hline \end{array}$$
$$\begin{array}{r} 8 \\ -\ 2 \\ \hline \end{array}$$

8.
$$\begin{array}{r} 12 \\ -\ 4 \\ \hline \end{array}$$
$$\begin{array}{r} 11 \\ -\ 5 \\ \hline \end{array}$$
$$\begin{array}{r} 9 \\ -\ 8 \\ \hline \end{array}$$
$$\begin{array}{r} 12 \\ -\ 7 \\ \hline \end{array}$$
$$\begin{array}{r} 8 \\ -\ 1 \\ \hline \end{array}$$
$$\begin{array}{r} 10 \\ -\ 4 \\ \hline \end{array}$$

9.
$$\begin{array}{r} 11 \\ -\ 9 \\ \hline \end{array}$$
$$\begin{array}{r} 10 \\ -\ 5 \\ \hline \end{array}$$
$$\begin{array}{r} 10 \\ -\ 8 \\ \hline \end{array}$$
$$\begin{array}{r} 12 \\ -\ 6 \\ \hline \end{array}$$
$$\begin{array}{r} 9 \\ -\ 4 \\ \hline \end{array}$$
$$\begin{array}{r} 11 \\ -\ 3 \\ \hline \end{array}$$

Use with Lesson 5-7, text pages 167–168.

Subtract. Show how you count back.

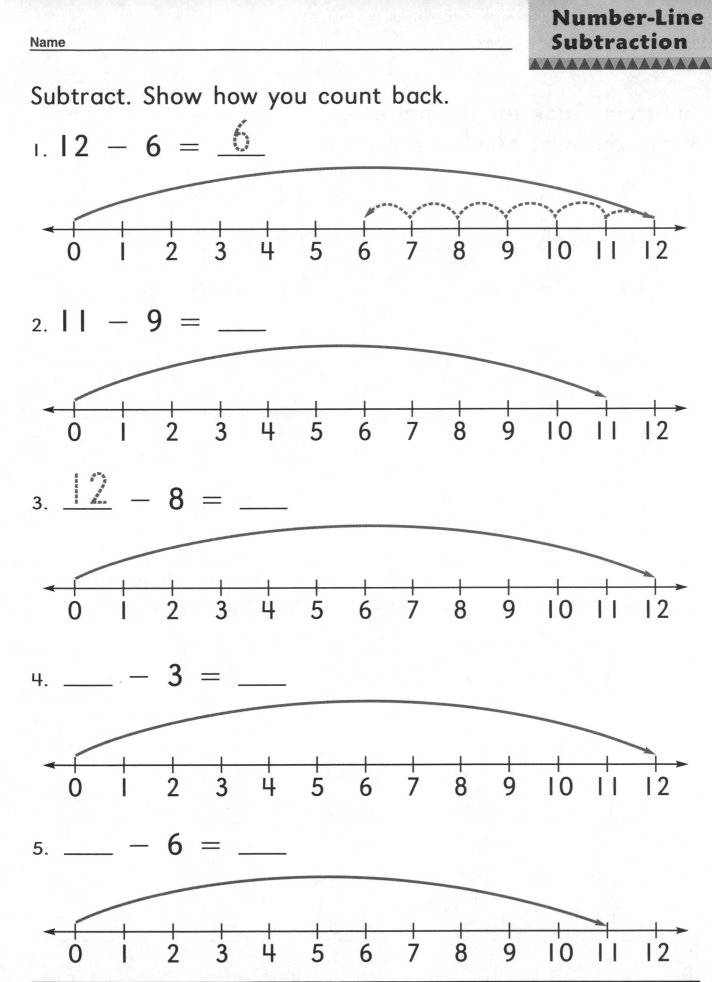

1. 12 − 6 = _6_

2. 11 − 9 = ___

3. _12_ − 8 = ___

4. ___ − 3 = ___

5. ___ − 6 = ___

Subtract. Look for the pattern.
Write the next number sentence.

1.
9 – 4 = __5__
10 – 5 = ___
11 – 6 = ___
__12__ – ___ = ___

2.
11 – 4 = ___
10 – 5 = ___
9 – 6 = ___
___ – ___ = ___

3.
12 – 6 = ___
12 – 5 = ___
12 – 4 = ___
___ – ___ = ___

4.
9 – 6 = ___
10 – 6 = ___
11 – 6 = ___
___ – ___ = ___

Subtract. Look for the pattern. Write the next fact.

5.
11 – 9 = 2   11 – 7   11 – 5   11 – 3   ___ –

6.
8 – 1   9 – 3   10 – 5   ___ –   ___ –

7.
9 – 5   8 – 5   7 – 5   ___ –   ___ –

Name _____

Subtract. Use a strategy you know.

1.	11 $-\ 4$ **7**	12 $-\ 5$	11 $-\ 7$	10¢ $-\ 9¢$ ¢	11¢ $-\ 5¢$ ¢	12¢ $-\ 3¢$ ¢
2.	12 $-\ 8$	10 $-\ 2$	11 $-\ 9$	10¢ $-\ 6¢$ ¢	10¢ $-\ 3¢$ ¢	12¢ $-\ 9¢$ ¢

3. $9 - 2 =$ _____ $11 - 6 =$ _____ $11 - 2 =$ _____

4. $10 - 5 =$ _____ $8 - 6 =$ _____ $10 - 5 =$ _____

Find the difference. Check by adding.

5.	10 $-\ 4$ **6**	**6** $+\ $**4**	6.	11 $-\ 6$	$+$	7.	12 $-\ 4$	$+$
8.	11 $-\ 8$	$+$	9.	12 $-\ 7$	$+$	10.	10 $-\ 8$	$+$
11.	9 $-\ 5$	$+$	12.	10 $-\ 7$	$+$	13.	12 $-\ 6$	$+$

Name _____

Complete these fact families.

1. 2 + 4 = _6_

   ___ ◯ ___ = ___

   ___ ◯ ___ = ___

   ___ ___ ___ = ___

2. 3 + 5 = ___

   ___ ◯ ___ = ___

   ___ ◯ ___ = ___

   ___ ___ ___ = ___

3. 4 + 3 = ___

   ___ ◯ ___ = ___

   ___ ◯ ___ = ___

   ___ ◯ ___ = ___

4. 2 + 6 = ___

   ___ ◯ ___ = ___

   ___ ◯ ___ = ___

   ___ ◯ ___ = ___

5. ⊕ 2 ◯ ___ ◯ ___ ◯ ___
       3
       —
       5

Add or subtract. Watch for + and −.

6.
$$
\begin{array}{cccccc}
10 & 6 & 11 & 3 & 4 & 12 \\
-\ 8 & +\ 5 & -\ 7 & +\ 9 & +\ 5 & -\ 6 \\
\hline
2 & & & & &
\end{array}
$$

7.
$$
\begin{array}{cccccc}
10 & 5 & 9 & 10 & 12 & 8 \\
-\ 7 & +\ 7 & +\ 2 & -\ 6 & -\ 3 & +\ 3 \\
\hline
\end{array}
$$

Use with Lesson 5-12, text pages 177–178.

1. Mai has 11 🪙.
David has 9 🪙.
Who has more?
How much more?

Mai | OOOOOOOOOOO
David | OOOOOOOOO

$\underline{\quad}$¢ $\bigcirc$ $\underline{\quad}$¢ = $\underline{\quad}$¢ $\underline{\qquad}$ has $\underline{\quad}$¢ more.

2. Tony won 12 🎗.
Sharon won 5 🎗.
Who has more 🎗?
How many more?

Tony | OOOOOOOOOOOO
Sharon | OOOOO

$\underline{\quad}$ $\bigcirc$ $\underline{\quad}$ = $\underline{\quad}$ $\underline{\qquad}$ has $\underline{\quad}$ 🎗 more.

3. Jody ate 10 🍪.
Mike ate 6 🍪.
Who ate fewer 🍪?
How many fewer?

Jody |
Mike |

$\underline{\quad}$ $\bigcirc$ $\underline{\quad}$ = $\underline{\quad}$ $\underline{\qquad}$ ate $\underline{\quad}$ fewer 🍪.

4. Flor has 12 🪙.
Luke has 8 🪙.
Who has fewer 🪙?
How many fewer?

Flor |
Luke |

$\underline{\quad}$ $\bigcirc$ $\underline{\quad}$ = $\underline{\quad}$ $\underline{\qquad}$ has $\underline{\quad}$ fewer 🪙.

Read ──────────▶ Model ──────▶ Think ─▶ Write

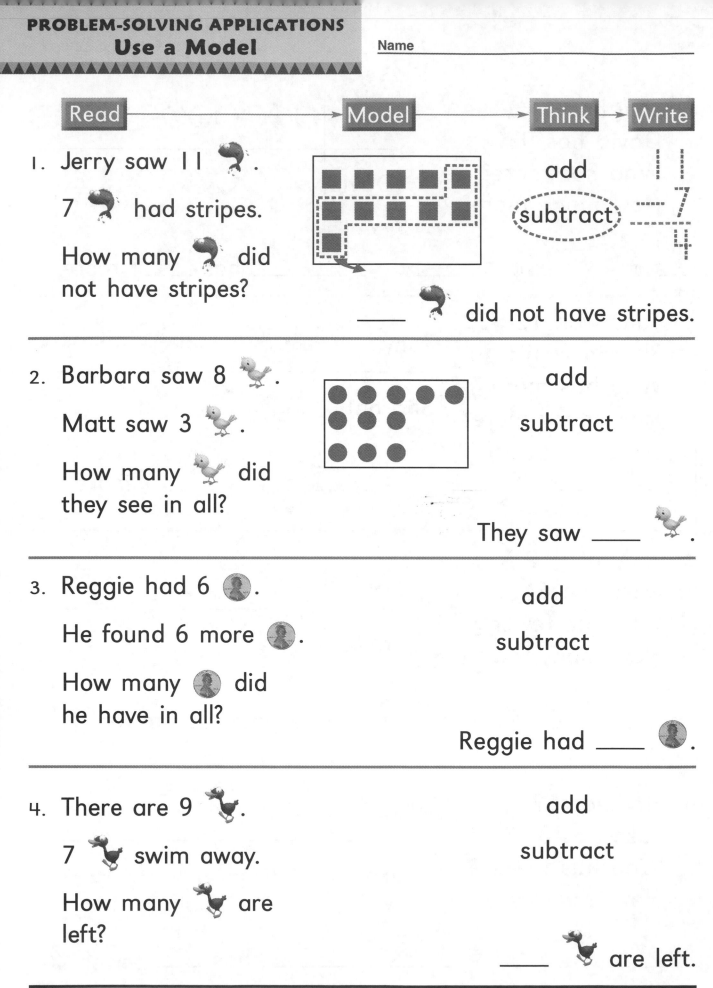

1. Jerry saw 11 🐟.

   7 🐟 had stripes.

   How many 🐟 did not have stripes?

   add

   (subtract)

   $\begin{array}{r} 11 \\ -\ 7 \\ \hline 4 \end{array}$

   _____ 🐟 did not have stripes.

2. Barbara saw 8 🐥.

   Matt saw 3 🐥.

   How many 🐥 did they see in all?

   add

   subtract

   They saw _____ 🐥.

3. Reggie had 6 🪙.

   He found 6 more 🪙.

   How many 🪙 did he have in all?

   add

   subtract

   Reggie had _____ 🪙.

4. There are 9 🦆.

   7 🦆 swim away.

   How many 🦆 are left?

   add

   subtract

   _____ 🦆 are left.

Use with Lesson 5-14, text pages 181–182.

Name _____

Find each sum or difference.

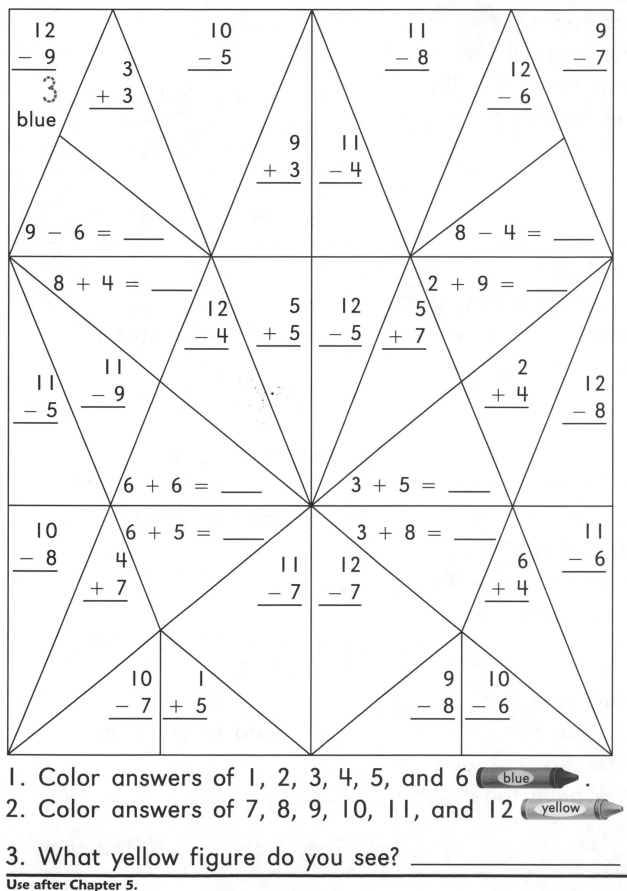

12
− 9
___
3
blue

3
+ 3
___

10
− 5
___

11
− 8
___

12
− 6
___

9
− 7
___

9
+ 3
___

11
− 4
___

9 − 6 = ___

8 − 4 = ___

8 + 4 = ___

2 + 9 = ___

12
− 4
___

5
+ 5
___

12
− 5
___

5
+ 7
___

11
− 5
___

11
− 9
___

2
+ 4
___

12
− 8
___

6 + 6 = ___

3 + 5 = ___

10
− 8
___

6 + 5 = ___

3 + 8 = ___

11
− 6
___

4
+ 7
___

11
− 7
___

12
− 7
___

6
+ 4
___

10
− 7
___

1
+ 5
___

9
− 8
___

10
− 6
___

1. Color answers of 1, 2, 3, 4, 5, and 6 blue.
2. Color answers of 7, 8, 9, 10, 11, and 12 yellow.

3. What yellow figure do you see? _____

Use a strategy you have learned.

1. Pat had 12 🪙.
She spent 8 🪙.
How many 🪙 did
she have left?

add or (subtract)

12 ⊝ 8 = ___

She had ___ 🪙 left.

2. Jim has 6¢.
Selma gave him 5 🪙.
How much does
he have now?

add or subtract

___¢ ◯ ___¢ = ___¢

He has ___¢.

3. Peg has 12 🎈.
Mary has 7 🎈.
Who has more 🎈?
How many more?

add or subtract

___ ◯ ___ = ___

_____ has ___ more 🎈.

4. Mark has 9 🎈.
Bill has 3 🎈.
How many 🎈 do
they have together?

add or subtract

___ ◯ ___ = ___

They have ___ 🎈 together.

56

Use after Chapter 5.

Name _____

Write how many.

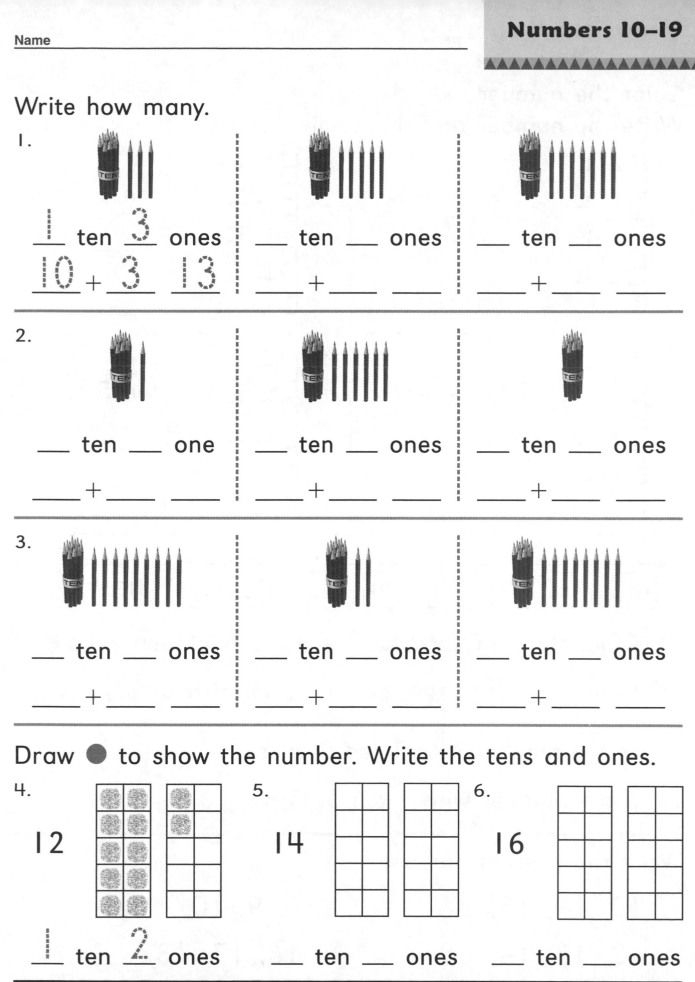

1.

_1_ ten _3_ ones

_10_ + _3_   _13_

___ ten ___ ones

___ + ___   ___

___ ten ___ ones

___ + ___   ___

2.

___ ten ___ one

___ + ___   ___

___ ten ___ ones

___ + ___   ___

___ ten ___ ones

___ + ___   ___

3.

___ ten ___ ones

___ + ___   ___

___ ten ___ ones

___ + ___   ___

___ ten ___ ones

___ + ___   ___

Draw ● to show the number. Write the tens and ones.

4.

12

_1_ ten _2_ ones

5.

14

___ ten ___ ones

6.

16

___ ten ___ ones

Name _____

Color the number.
Write the number and the number word.

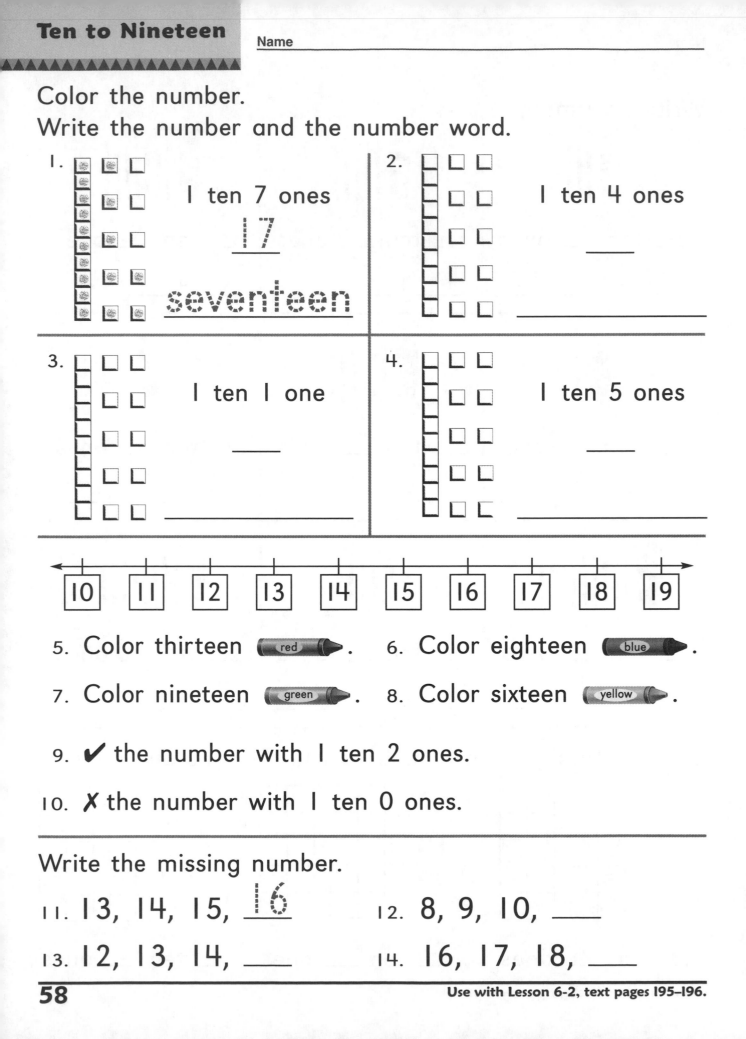

1. 1 ten 7 ones

17

seventeen

2. 1 ten 4 ones

____

_____

3. 1 ten 1 one

____

_____

4. 1 ten 5 ones

____

_____

10 11 12 13 14 15 16 17 18 19

5. Color thirteen ⬤ red ▶. 6. Color eighteen ⬤ blue ▶.

7. Color nineteen ⬤ green ▶. 8. Color sixteen ⬤ yellow ▶.

9. ✔ the number with 1 ten 2 ones.

10. ✗ the number with 1 ten 0 ones.

Write the missing number.

11. 13, 14, 15, 16   12. 8, 9, 10, ___

13. 12, 13, 14, ___   14. 16, 17, 18, ___

Use with Lesson 6-2, text pages 195–196.

Name _____

Count by 10s. Write the missing numbers.

1. 10, 20, __30__, 40, ___, 60, 70, ___, 90

2. 20, ___, ___, ___, ___, ___, 80, ___

3. 30, ___, ___, ___, 70, ___, 90

Write the number.

4.		
2 tens 6 ones	2 tens 3 ones	3 tens 1 one
__26__	___	___
5.		
3 tens 6 ones	2 tens 2 ones	2 tens 4 ones
___	___	___
6.		
2 tens 5 ones	3 tens 0 ones	3 tens 8 ones
___	___	___
7.		
3 tens 4 ones	3 tens 7 ones	2 tens 9 ones
___	___	___
8.		
2 tens 0 ones	2 tens 1 one	3 tens 3 ones
___	___	___

Use with Lessons 6-3 and 6-4, text pages 197–200.

Name _____

Write the place value.

1. 42 = __4__ tens __2__ ones   58 = ___ tens ___ ones

2. 46 = ___ tens ___ ones   40 = ___ tens ___ ones

3. 57 = ___ tens ___ ones   44 = ___ tens ___ ones

Write the number word.

4. 52 = fifty-_____   43 = forty-_____

5. 56 = fifty-_____   48 = forty-_____

Write the number.

6. 4 tens 5 ones = __45__

   4 tens 9 ones = ____

   5 tens 0 ones = ____

   5 tens 2 ones = ____

   4 tens 8 ones = ____

   4 tens 1 one = ____

   4 tens 2 ones = ____

   5 tens 9 ones = ____

   5 tens 5 ones = ____

7. 5 tens 1 one = ____

   4 tens 3 ones = ____

   4 tens 7 ones = ____

   5 tens 3 ones = ____

   5 tens 0 ones = ____

   5 tens 6 ones = ____

   4 tens 4 ones = ____

   4 tens 9 ones = ____

   5 tens 4 ones = ____

Use with Lesson 6-5, text pages 201–202.

Name _____

Write the number and the place value.

Number Word	Number	Place Value
1. seventeen	17	1 ten 7 ones
2. thirty-one	___	___ tens ___ one
3. forty	___	___ tens ___ ones
4. fifty-six	___	___ tens ___ ones
5. twenty-eight	___	___ tens ___ ones
6. fifty-nine	___	___ tens ___ ones

Write the missing numbers.

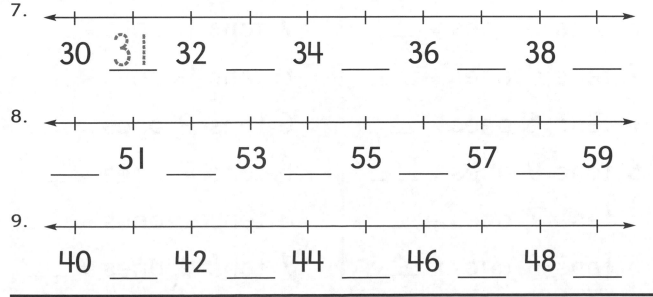

7. 30   31   32   ___   34   ___   36   ___   38   ___

8. ___   51   ___   53   ___   55   ___   57   ___   59

9. 40   ___   42   ___   44   ___   46   ___   48   ___

Name _____

Write the place value.

1. 79 = _7_ tens _9_ ones   66 = ___ tens ___ ones

2. 73 = ___ tens ___ ones   78 = ___ tens ___ ones

3. 60 = ___ tens ___ ones   70 = ___ tens ___ ones

Write the number word.

4. 72 = seventy-_____   67 = sixty-_____

5. 64 = sixty-_____   71 = seventy-_____

Write the number.

6. 6 tens 1 one = _61_   7. 6 tens 3 ones = ___

   7 tens 5 ones = ___   6 tens 2 ones = ___

   6 tens 8 ones = ___   7 tens 7 ones = ___

   6 tens 5 ones = ___   7 tens 0 ones = ___

   7 tens 0 ones = ___   6 tens 4 ones = ___

   7 tens 6 ones = ___   6 tens 9 ones = ___

   6 tens 9 ones = ___   6 tens 7 ones = ___

   7 tens 7 ones = ___   6 tens 2 ones = ___

   6 tens 3 ones = ___   7 tens 4 ones = ___

Use with Lesson 6-7, text pages 205–206.

Name _____

Write the place value.

1. 89 = __8__ tens __9__ ones   83 = ____ tens ____ ones

2. 93 = ____ tens ____ ones   87 = ____ tens ____ ones

3. 86 = ____ tens ____ ones   88 = ____ tens ____ ones

Write the number word.

4. 99 = ninety-_____   85 = eighty-_____

5. 94 = ninety-_____   81 = eighty-_____

Write the number.

6. 9 tens 5 ones = __95__   7. 9 tens 2 ones = ____

   9 tens 1 one = ____   8 tens 4 ones = ____

   8 tens 2 ones = ____   9 tens 0 ones = ____

   8 tens 0 ones = ____   8 tens 6 ones = ____

   9 tens 8 ones = ____   8 tens 5 ones = ____

   9 tens 6 ones = ____   9 tens 9 ones = ____

   9 tens 2 ones = ____   9 tens 7 ones = ____

   8 tens 8 ones = ____   8 tens 1 one = ____

   8 tens 4 ones = ____   8 tens 8 ones = ____

Name _____

Write the missing numbers.

1.
60 __61__ 62 ___ 64 ___ 66 ___ 68 ___ 70

2.
___ 51 52 53 ___ 55 ___ 57 58 59 ___

3.
90 ___ 92 93 ___ 95 96 ___ 98 ___ 100

4.
70 ___ 72 ___ 74 75 76 ___ 78 ___ 80

Complete the place value. Color the number.

5. sixty-five ones

__65__ ones = __6__ tens __5__ ones

6. fifty-six ones

____ ones = __ tens __ ones

7. seventy-eight ones

____ ones = __ tens __ ones

8. eighty-seven ones

____ ones = __ tens __ ones

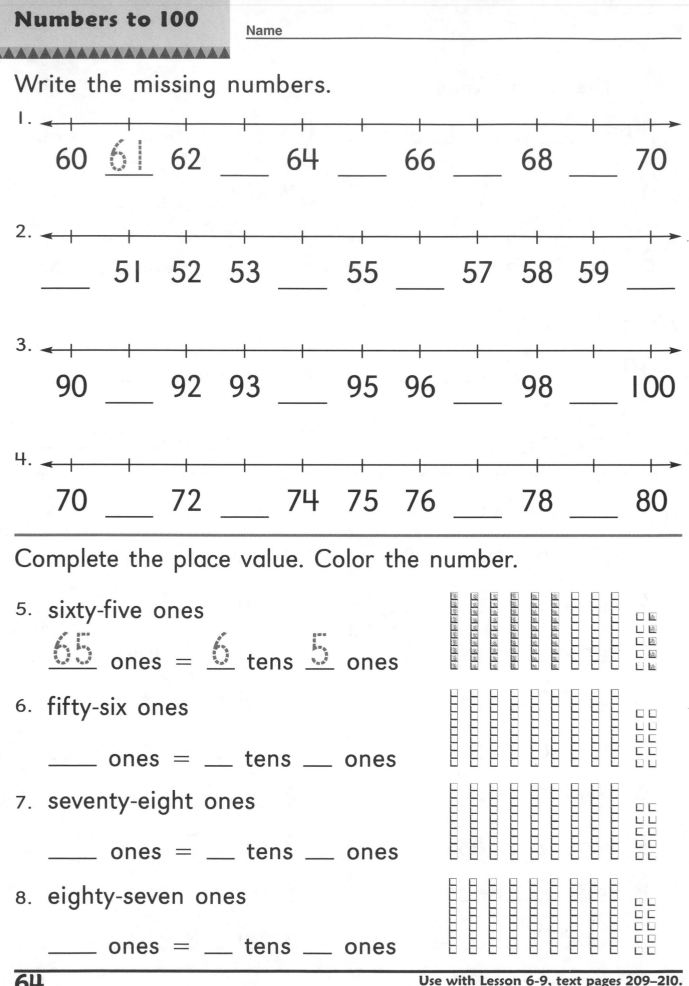

Use with Lesson 6-9, text pages 209–210.

1. Write the missing numbers.

1	2	3	4	5	6				10
11			14	15	16				20
		24					28	29	30
	32			35			38		40
	42		44				48		50
			54		56	57			60
61	62	63	64	65	66	67		69	70
			74						
	82	83	84		86				
91							98	99	

Write the missing numbers.

2. 30, 40, 50, ___, ___

3. 57, 56, 55, ___, ___

4. 73, 74, 75, ___, ___

5. 93, 83, 73, ___, ___

Use the hundred chart above.
Write the missing numbers.

6.
1		3		5
11	12	13	14	
				25

7.
71	72			75
81				
91				95

Count On and Count Back; Before, Between, After

Name_____

Write the number that comes just after.

1. 82, 83, __84__

2. 11, 12, ____

3. 45, 46, ____

4. 34, 35, ____

5. 57, 58, ____

6. 79, 80, ____

Write the number that comes just before.

7. __81__, 82, 83

8. ____, 29, 30

9. ____, 67, 68

10. ____, 94, 95

11. ____, 18, 19

12. ____, 61, 62

Write the number that comes between.

| | Between | | | | Between | |
| Before | | After | | Before | | After |

13. 15, __16__, 17

14. 76, ____, 78

15. 40, ____, 42

16. 39, ____, 41

Write the numbers that each number is between.

| Before | | After | | Before | | After |

17. __21__, 22, __23__

18. ____, 68, ____

19. ____, 87, ____

20. ____, 71, ____

Use with Lessons 6-11, 6-12, and 6-13, text pages 213–215.

Name _____

Compare. Ring < or >.

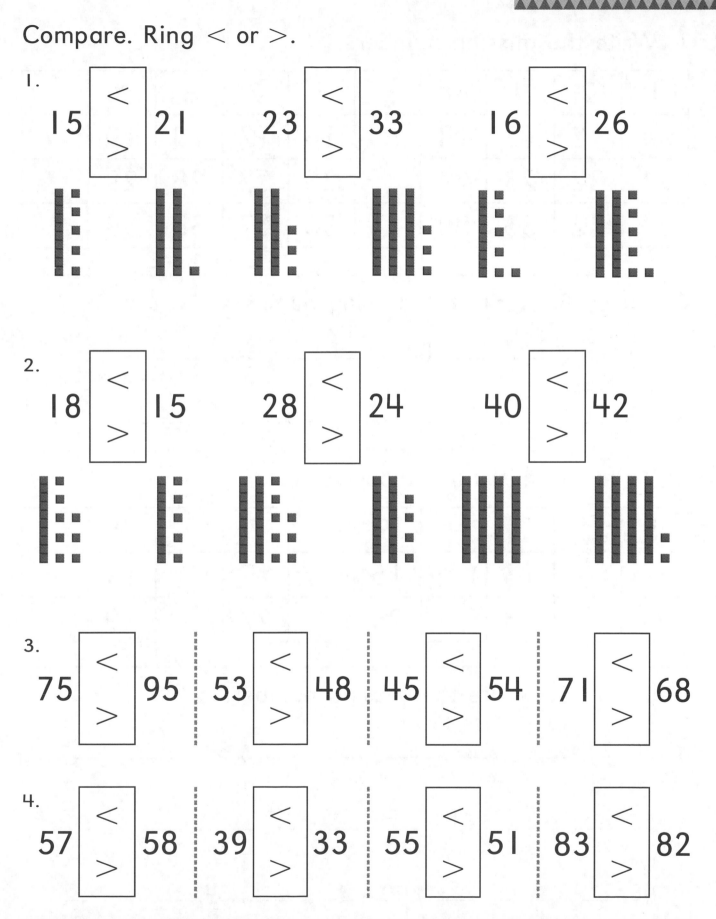

1.

15 < > 21   23 < > 33   16 < > 26

2.

18 < > 15   28 < > 24   40 < > 42

3.

75 < > 95   53 < > 48   45 < > 54   71 < > 68

4.

57 < > 58   39 < > 33   55 < > 51   83 < > 82

Name _____

1. Write the missing numbers.

1	2	3	4	5	6	7	8	9	
11	12	13	14		16	17	18	19	
21	22	23	24		26	27	28	29	
31	32	33	34		36	37	38	39	

Count by 5s. Write the missing numbers.

2. 60, 65, _70_, ____, 80, ____, ____, 95, ____

3. 25, ____, ____, 40, ____, ____, 55, ____, ____

4. Write the missing numbers.

1	2	3	4	5		7		9	
11		13		15		17		19	
21		23		25		27		29	

Count by 2s. Write the missing numbers.

5. 70, 72, _74_, ____, ____, ____, ____, 84, ____, ____

6. 55, ____, ____, 61, ____, ____, 67, ____, ____, ____

7. 48, ____, ____, ____, 40, ____, ____, 34, ____, ____

Use with Lessons 6-15 and 6-16, text pages 219–222.

Ring groups of 5. Write the missing numbers.

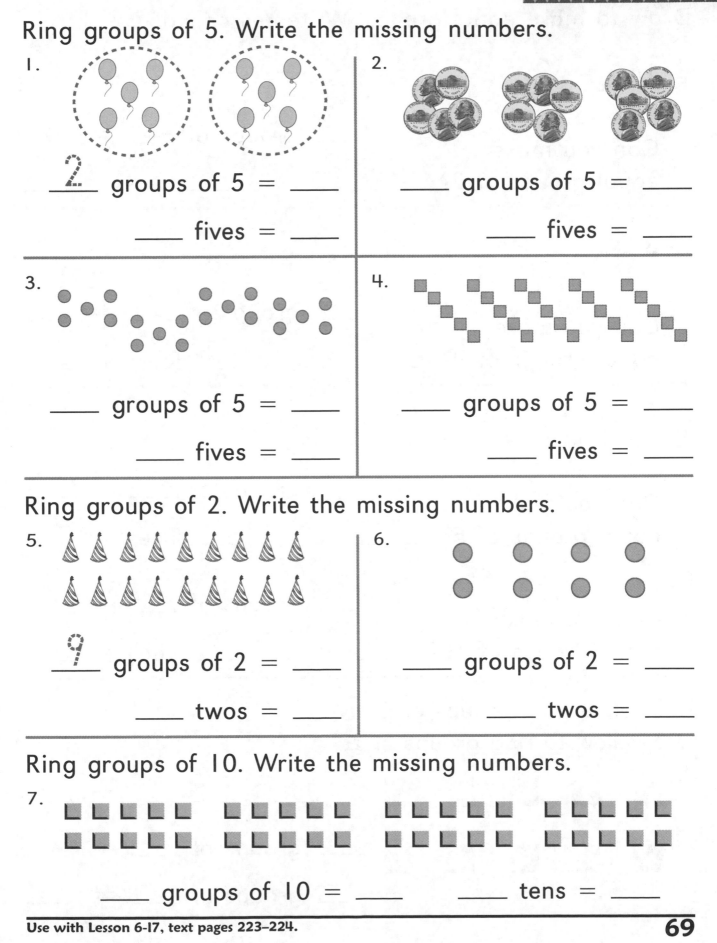

1. _2_ groups of 5 = ____

   ____ fives = ____

2. ____ groups of 5 = ____

   ____ fives = ____

3. ____ groups of 5 = ____

   ____ fives = ____

4. ____ groups of 5 = ____

   ____ fives = ____

Ring groups of 2. Write the missing numbers.

5. _9_ groups of 2 = ____

   ____ twos = ____

6. ____ groups of 2 = ____

   ____ twos = ____

Ring groups of 10. Write the missing numbers.

7. ____ groups of 10 = ____   ____ tens = ____

Use with Lesson 6-17, text pages 223–224.

Name _____

Draw to make equal groups. Write Yes or No.

1. 14 🔵 ⭕⭕⭕⭕⭕⭕⭕
      ⭕⭕⭕⭕⭕⭕⭕

Can you make
equal groups of 2? __Yes__

__7__ groups of __2__ = 14

__7__ twos = ____

2. 8 🪀

Can you make
equal groups of 4? _____

____ groups of ____ = ____

____ fours = ____

3. 20 🎈

Can you make
equal groups of 5? _____

____ groups of ____ = ____

____ fives = ____

Can you make
equal groups of 10? _____

____ groups of ____ = ____

____ tens = ____

Use red to ring groups of 5.
Use yellow to ring groups of 2.

4.

⬜ ⬜ ⬜ ⬜ ⬜

⬜ ⬜ ⬜ ⬜ ⬜

____ groups of ____ = ____

____ groups of ____ = ____

Use with Lesson 0-0, text pages 000–000.

Name _____

Share. Tally. Write how many in each group.

1. 2 friends share 10 🪙 equally.

   Julia   Pat

   How many 👤 does each get?

   __5__ 👤 each

2. Three friends share 12 baseball cards equally.

   Gerard   Lucy   Joe

   __4__ cards each

3. Four friends share 8 marbles equally.

   Brita   James   Ann   Sam

   _____ marbles each

4. Five friends share 10 toy cars equally.

   Noah   Ben   June   Steve   Zeke

   _____ toy cars each

Ring to show equal groups.
Write how many groups.

1. Stu put 8 dimes into groups
of 2. How many groups
of two did Stu make?

   __8__ = __4__ groups of two

2. Flo put 15 bows into groups
of five. How many groups
of five did she make?

   ____ = ____ groups of five

Draw ◻ to trade.
Write how many equal groups.

3. 1 ▱▱▱▱▱▱▱▱▱▱ 2 ◻
   4 in each group

   __12__ = __3__ groups of four

4. 1 ▱▱▱▱▱▱▱▱▱▱ 0 ◻
   2 in each group

   ____ = ____ groups of two

5. 2 ▰▰▰▰▰▰▰▰▰▰ 0 ◻
   ▰▰▰▰▰▰▰▰▰▰
   2 in each group

   ____ = ____ groups of two

6. 2 ▰▰▰▰▰▰▰▰▰▰ 4 ◻
   ▰▰▰▰▰▰▰▰▰▰
   4 in each group

   ____ = ____ groups of four

Use with Lesson 6-19, text pages 227–230.

Name _____

Use these numbers.

13	29	44	50	55	64	72	87	90

1. I am an odd number.
   I am greater than 40 but less than 80.

   Which numbers are odd?   _13_, ____, ____, ____

   Which of these are greater than 40?   ____, ____

   Which is less than 80?   ____

   I am ____.

2. I am an even number between 20 and 90.
   You say my name when you count by 5s.

   Which numbers are even?   ____, ____, ____, ____, ____

   Which do you say when you count by 5s?   ____, ____

   Which is between 20 and 90?   ____

   I am ____.

3. You say my name when you count by 2s.
   I am greater than 10 but less than 50.

   Which do you say when you count by 2s?

   ____, ____, ____, ____, ____

   Which is between 10 and 50?   ____

   I am ____.

Use with Lesson 6-21, text pages 231–232.

1. **Read** Marc counts from 58 through 63.
Carrie counts from 64 through 69.
What odd numbers did they count?

Think List the numbers that
Marc and Carrie counted.

Ring the odd numbers.

58 , ____, ____, ____, ____, 63

64 , ____, ____, ____, ____, 69

Write Write the numbers with a ring.

Check _____ are odd.

2. **Read** George put the numbers 89
through 99 in a box. Marie
took out 91, 93, 94, 96, and 99.
What numbers are in the box?

Think List the numbers in the box.
✗ the numbers Marie took out.

89 , ____, ____, ____, ____, 94 ,

95 , ____, ____, ____, 99

Write Write the numbers still in the box.

Check 89 , ____, ____, ____, ____, ____

Use with Lesson 6-22, text pages 233–234.

Name ____________________

Write how much.

1.

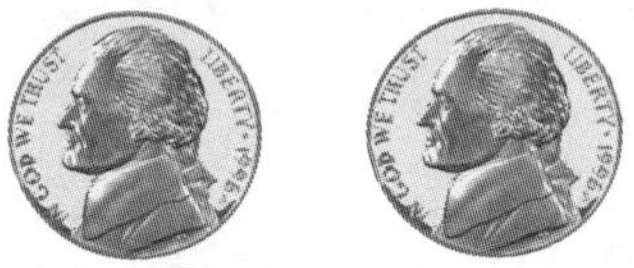

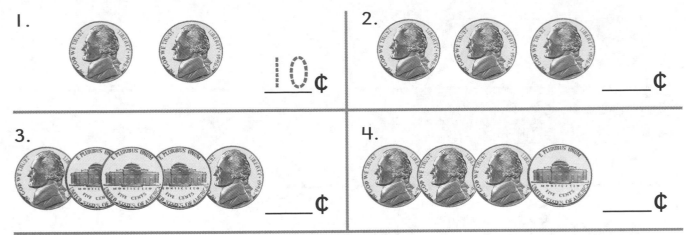

10¢

2.

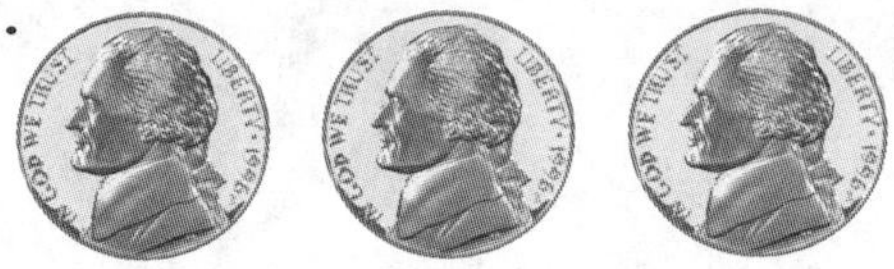

_____¢

3.

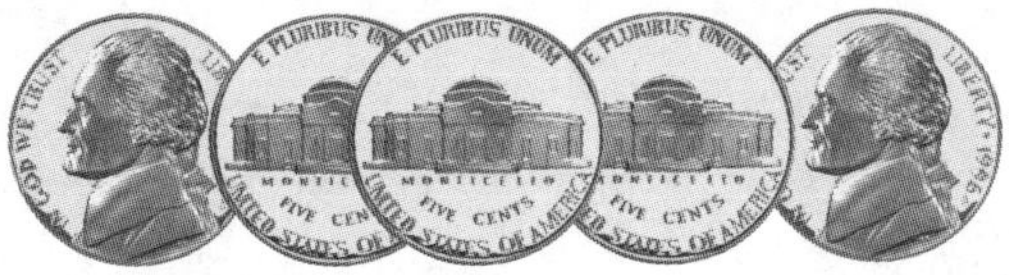

_____¢

4.

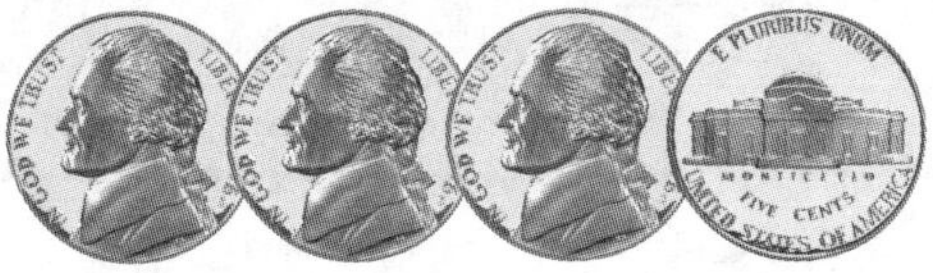

_____¢

Ring 30¢.

5.

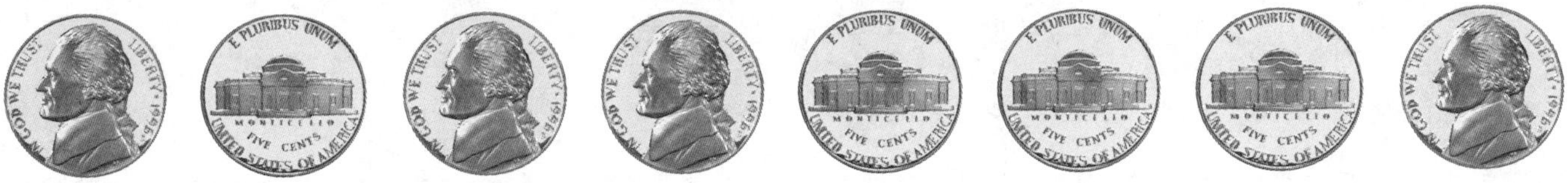

Write how much.

6.

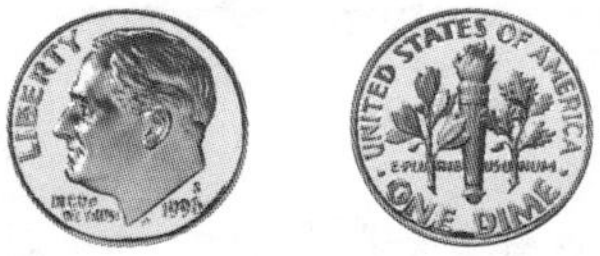

20¢

7.

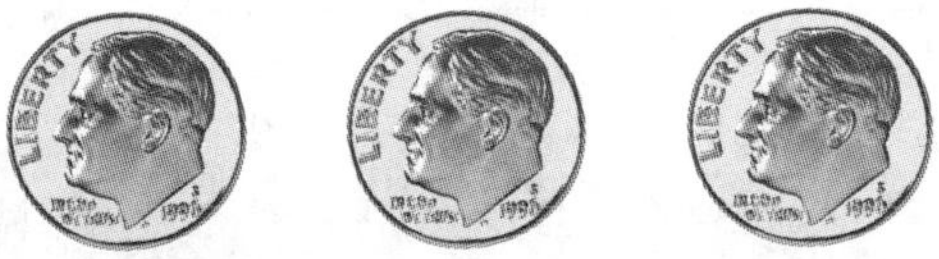

_____¢

8.

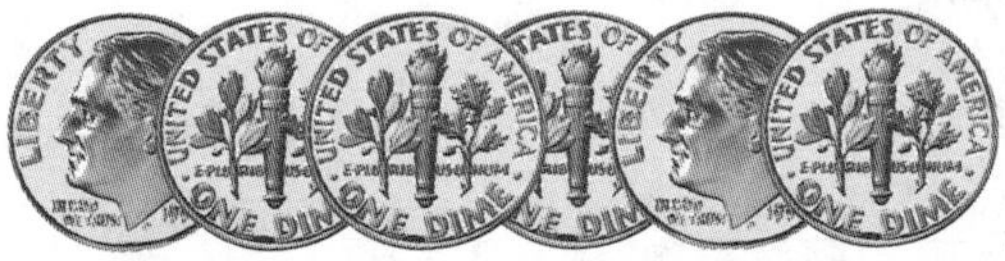

_____¢

9.

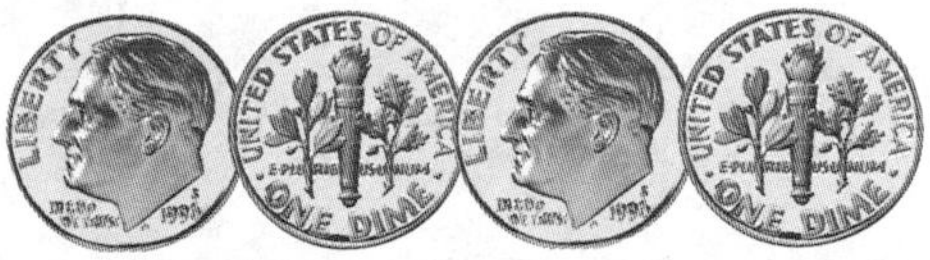

_____¢

Ring 50¢.

10.

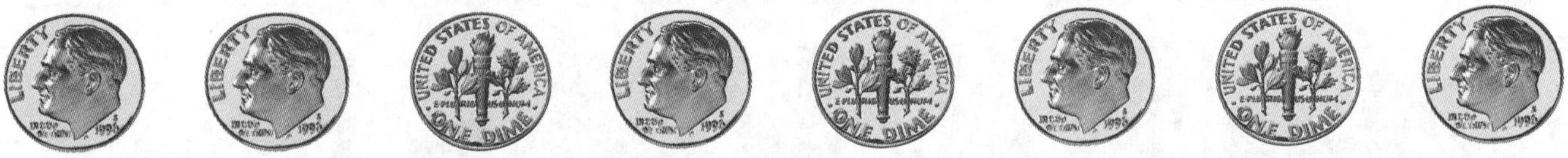

Write how much.

1. 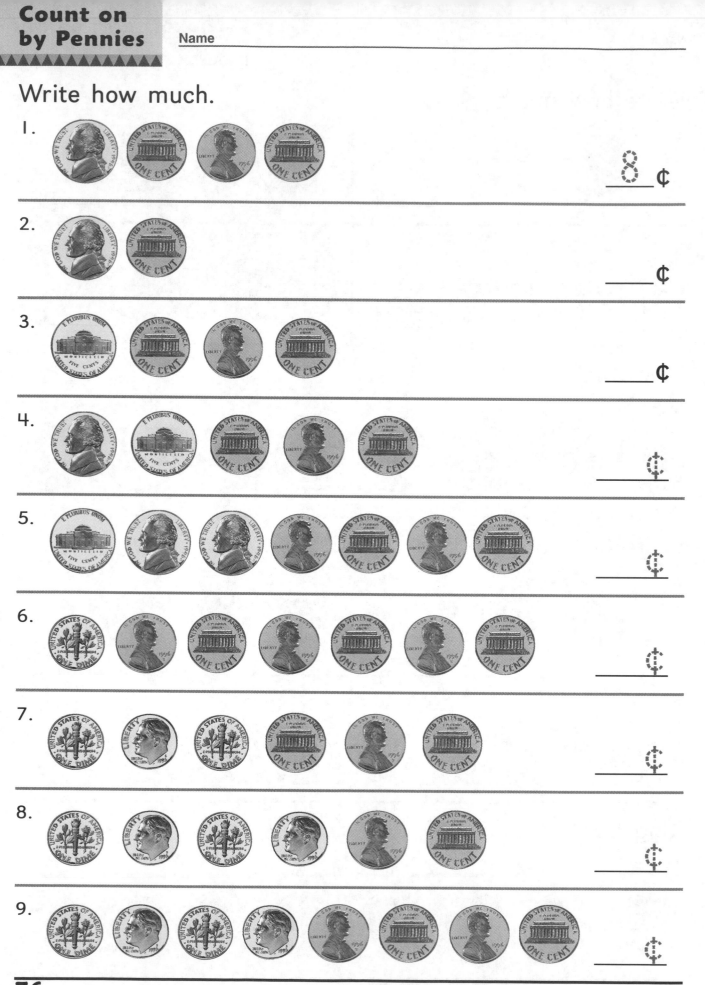 ___8___ ¢

2. _____ ¢

3. _____ ¢

4. _____ ¢

5. _____ ¢

6. _____ ¢

7. _____ ¢

8. _____ ¢

9. _____ ¢

Use with Lesson 7-3, text pages 245–246.

Name

Write how much.

1.

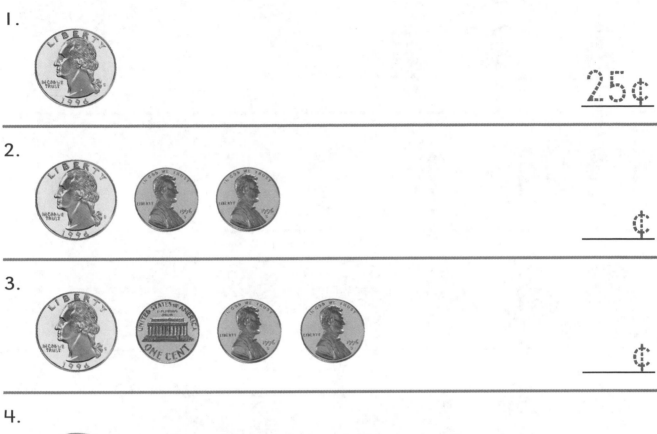

25¢

2.

_____ ¢

3.

_____ ¢

4.

_____ ¢

Ring the amount.

5.

29¢

6.

32¢

Write how much.

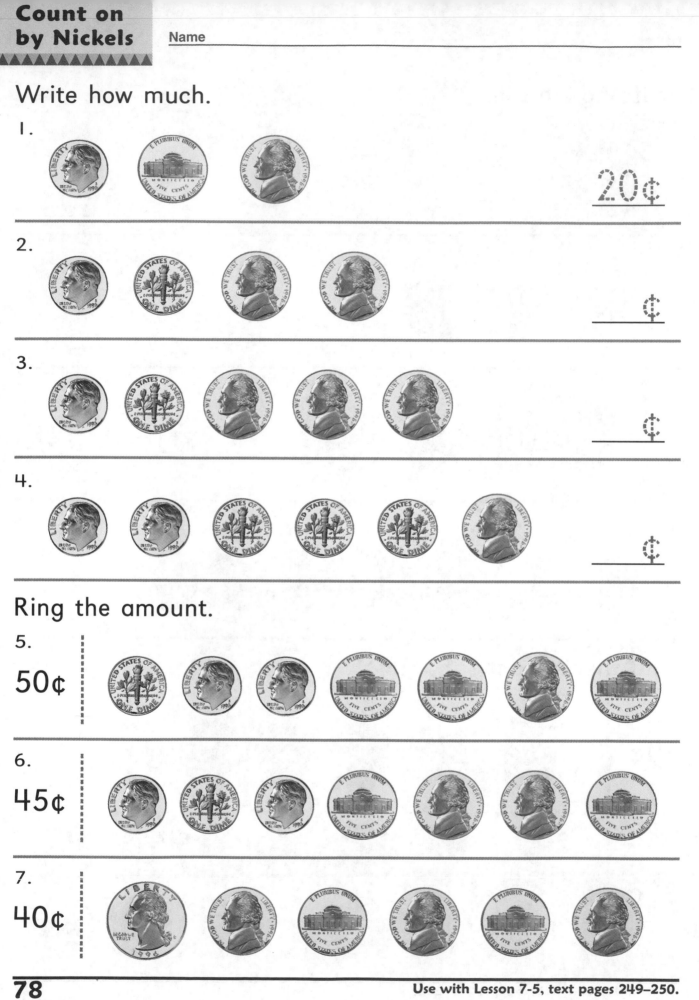

1. 20¢

2. ____ ¢

3. ____ ¢

4. ____ ¢

Ring the amount.

5. 50¢

6. 45¢

7. 40¢

Use with Lesson 7-5, text pages 249–250.

Name

Write how much.

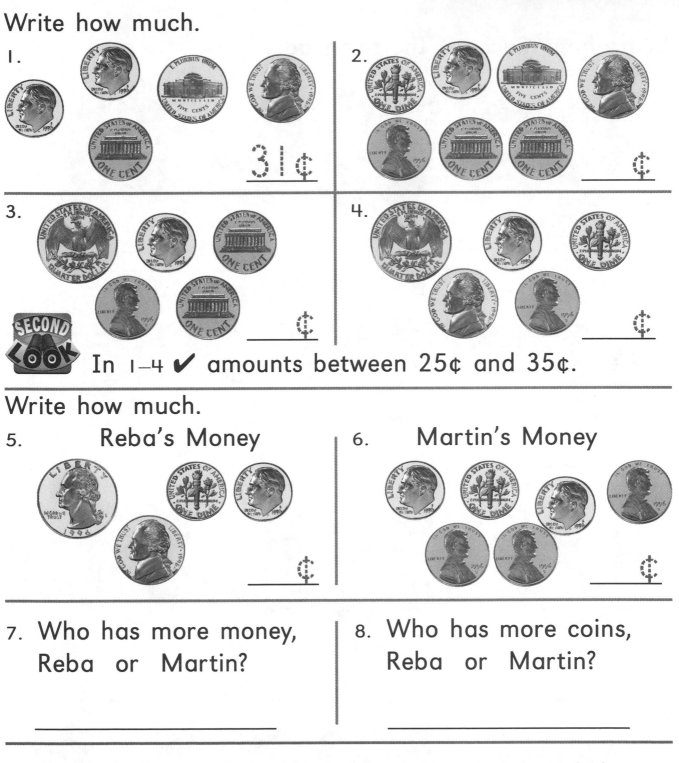

1. 31¢

2. _____ ¢

3. _____ ¢

4. _____ ¢

In 1–4 ✔ amounts between 25¢ and 35¢.

Write how much.

5. Reba's Money _____ ¢

6. Martin's Money _____ ¢

7. Who has more money, Reba or Martin?

_____

8. Who has more coins, Reba or Martin?

_____

9. Draw 2 dimes, 1 quarter, 3 pennies and 1 nickel in the easiest order to count on.

_____ ¢

Use with Lesson 7-6, text pages 251–252.

Write how much. ✔ the fair trade.

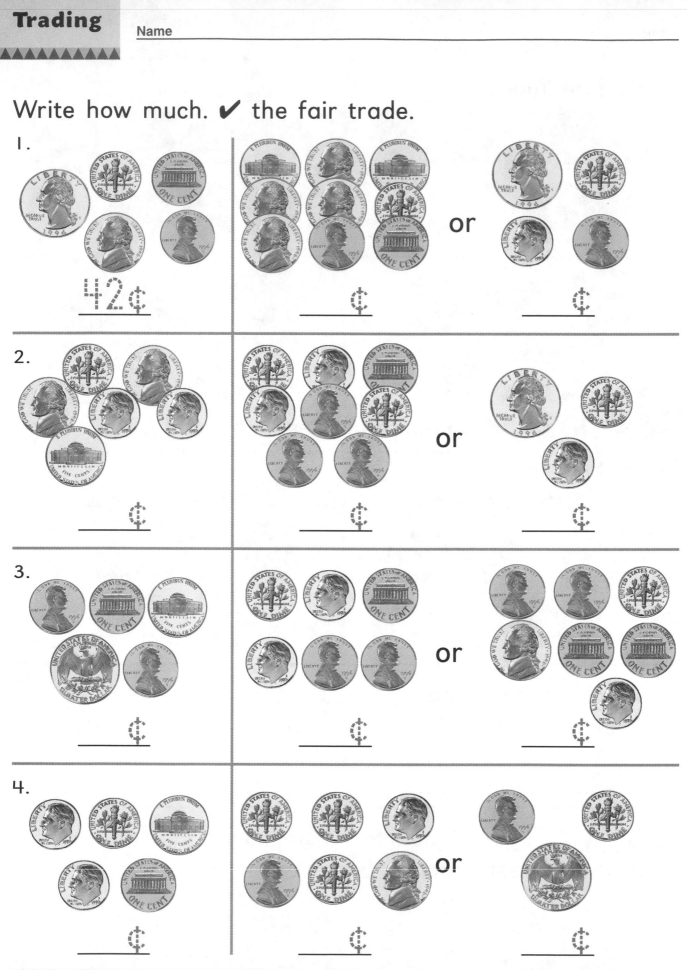

1.

42¢ _____

_____ ¢   or   _____ ¢

2.

_____ ¢

_____ ¢   or   _____ ¢

3.

_____ ¢

_____ ¢   or   _____ ¢

4.

_____ ¢

_____ ¢   or   _____ ¢

Use with Lesson 7-7, text pages 253–254.

Name

Write the amount. Match.

1.  30¢ 28¢

2. _____¢ 30¢

3. _____¢ 45¢

Trade and compare to make the amounts equal.

4. Jian has

 _____   _____

Sally has

 _____

Jian needs _____ _____.

Write the time in 2 ways.

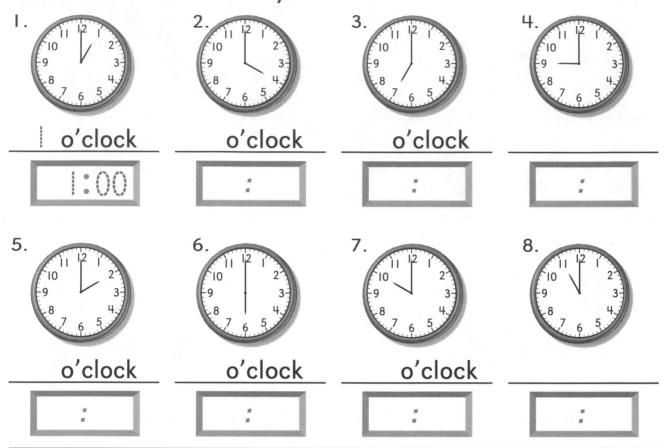

1. _ı_ o'clock

1:00

2. ___ o'clock

:

3. ___ o'clock

:

4. ___

:

5. ___ o'clock

:

6. ___ o'clock

:

7. ___ o'clock

:

8. ___

:

Show the time. Draw the hour hand.

9. 3 o'clock   5 o'clock   8 o'clock   12 o'clock

10.   7:00   2:00   4:00   1:00

Use with Lesson 7-10, text pages 259–260.

Name _____

Write the time.

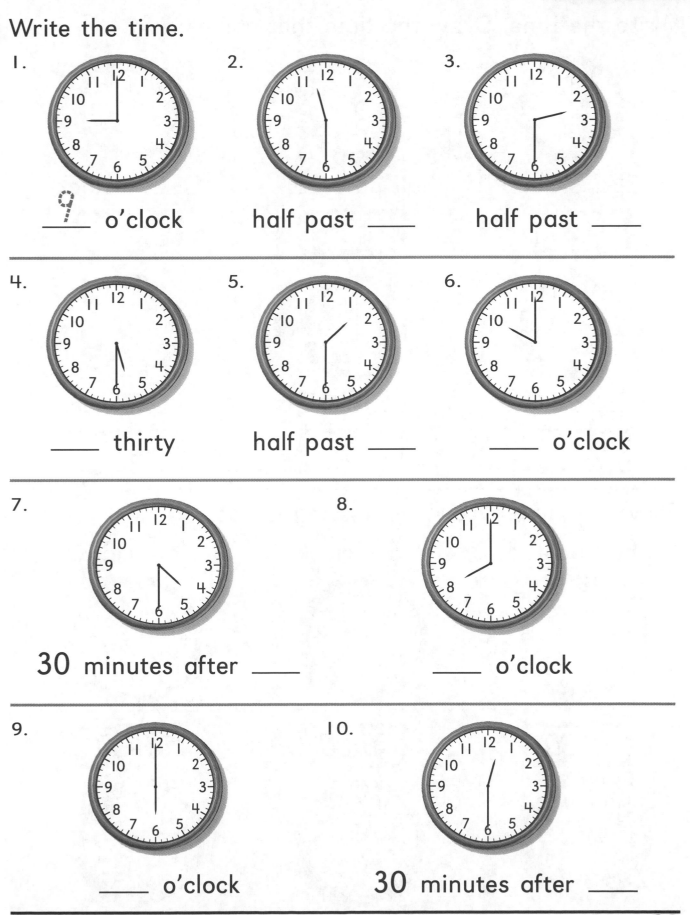

1. _____ o'clock

2. half past _____

3. half past _____

4. _____ thirty

5. half past _____

6. _____ o'clock

7. **30** minutes after _____

8. _____ o'clock

9. _____ o'clock

10. **30** minutes after _____

Write the time. Draw the time that comes next.

1.

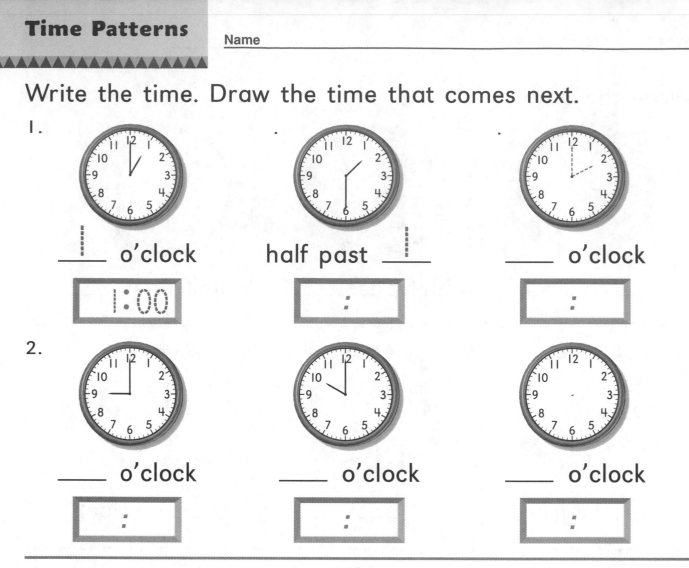

_____ o'clock   half past _____   _____ o'clock

| 1:00 | | : | | : |

2.

_____ o'clock   _____ o'clock   _____ o'clock

| : | | : | | : |

Show the time. Draw the missing hands.

3. half past **3**   **3** o'clock   **4** o'clock

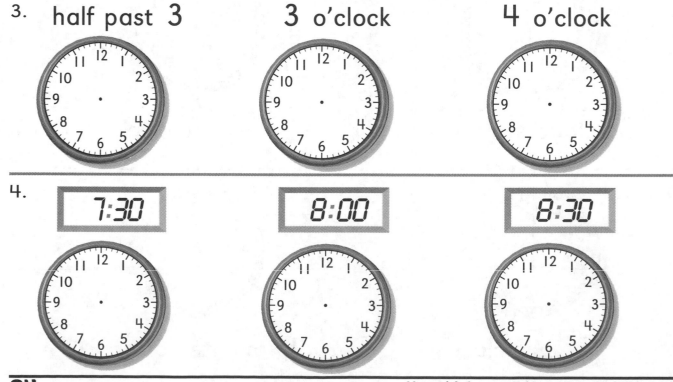

4. | 7:30 | | 8:00 | | 8:30 |

Use with Lesson 7-12, text pages 263–264.

Name _____

Draw the time.

1. The game begins at 10:30.   The game ends at 11:30.

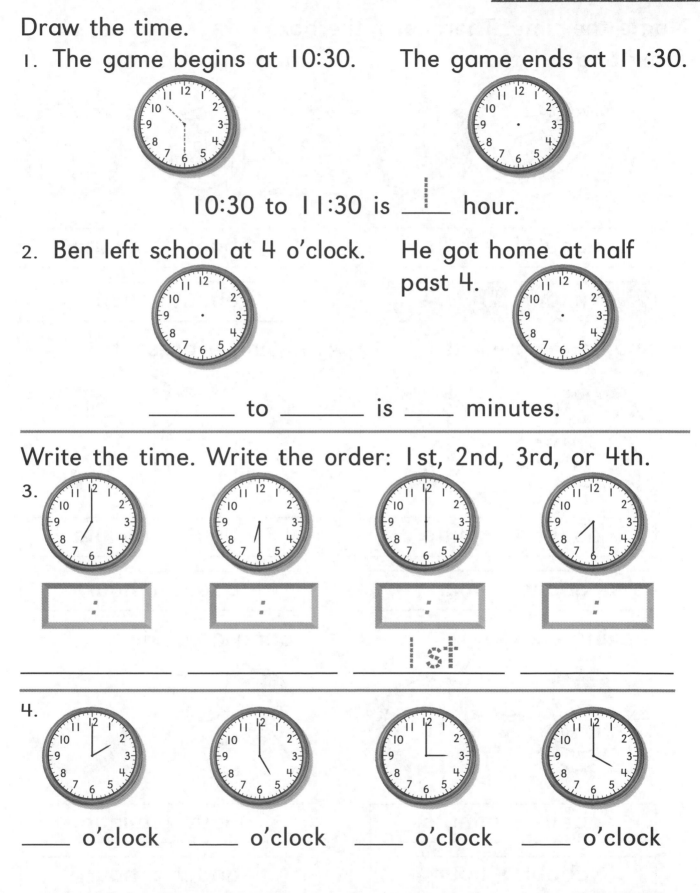

10:30 to 11:30 is ____ hour.

2. Ben left school at 4 o'clock.   He got home at half past 4.

_____ to _____ is ___ minutes.

Write the time. Write the order: 1st, 2nd, 3rd, or 4th.

3.

[:]   [:]   [:]   [:]

_____ _____ __1st__ _____

4.

___ o'clock   ___ o'clock   ___ o'clock   ___ o'clock

_____ _____ _____ _____

Name the time. Then color the box.

1. writing a letter

about I minute

about I hour

2. mailing a letter

about I minute

about I hour

3. opening a present

about I minute

about I hour

4. buying a present

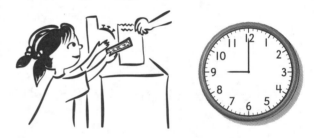

about I minute

about I hour

5. making cookies

about I minute

about I hour

6. eating a cookie

about I minute

about I hour

Use with Lesson 7-14, text pages 267–268.

Name _____

Sunday	Monday	Tuesday	Wednesday	Thursday	Friday	Saturday
			1	2	3	4
5	6	7	8	9	10	11
12	13	14	15	16	17	18
19	20	21	22	23	24	25
26	27	28	29	30		

1. Write the day of the week for the first day of this month.  Wednesday

2. Write the day of the week for the last day of this month. _____

3. Write the number of days in this month. _____

4. What day of the week is each date?

   6 _____ 17 _____ 26 _____

   23 _____ 14 _____ 11 _____

5. Color all the Wednesdays.

6. Write the number of Wednesdays. _____

7. Write the dates of all the Wednesdays. _____

Use with Lesson 7-15, text pages 269–270.

Draw the time on the clock.
Then ring the time that makes sense.

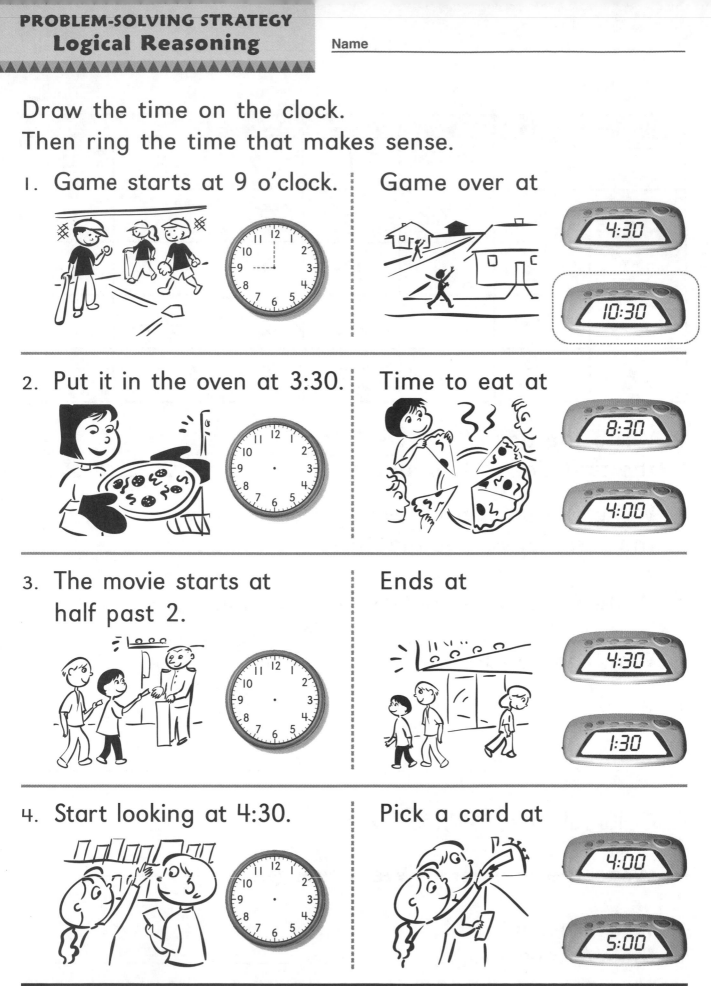

1. Game starts at 9 o'clock.   Game over at

4:30

10:30

2. Put it in the oven at 3:30.   Time to eat at

8:30

4:00

3. The movie starts at
   half past 2.   Ends at

4:30

1:30

4. Start looking at 4:30.   Pick a card at

4:00

5:00

Use with Lesson 7-16, text pages 271–272.

1. **Read** I need a dozen eggs for the picnic.
There are only 7 in the box.
How many eggs do I still need?

   Think __12__ eggs in a dozen.

   Write $12 \ominus 7 = 5$

   Check I need __5__ eggs.

---

2. **Read** Dad gave Mom a dozen roses.
6 were white. The other roses were red.
How many were red?

   Think ____ roses in a dozen

   Write ___ ◯ ___ = ___

   Check ____ roses were red.

---

3. **Read** Alice needs 2 nickels for a pretzel.
She finds 2 nickels and 2 pennies.
How much money is left after she
buys a pretzel?

   Think 2 nickels are ____¢

   2 nickels and two pennies are ____¢

   Write ___¢ ◯ ___¢ = ___¢

   Check ___¢ is left.

✔ each open figure.

1.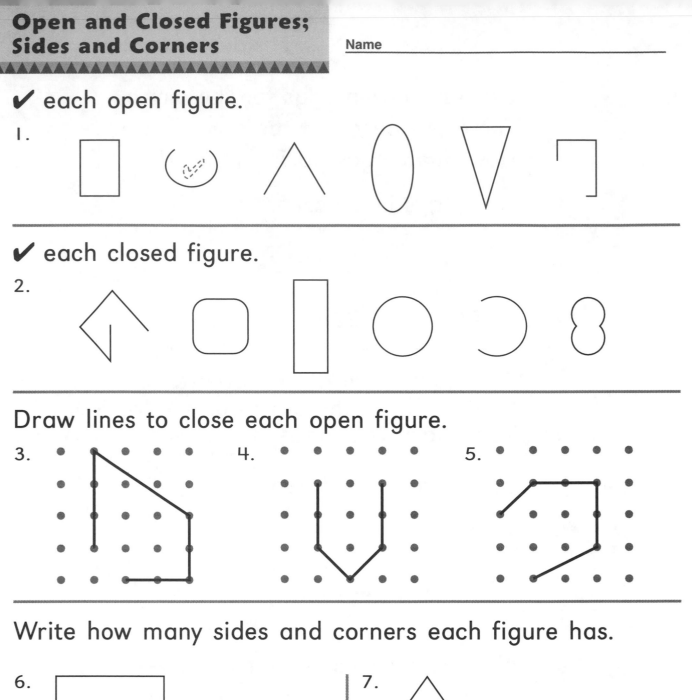

✔ each closed figure.

2.

Draw lines to close each open figure.

3.   4.   5.

Write how many sides and corners each figure has.

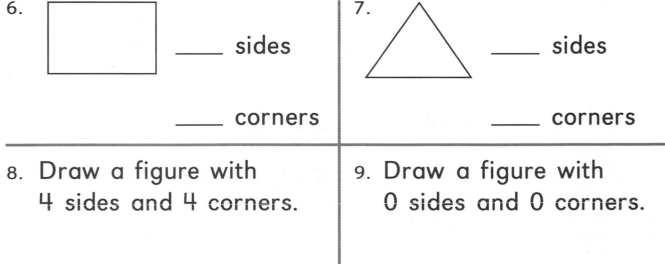

6. _____ sides

_____ corners

7. _____ sides

_____ corners

8. Draw a figure with
4 sides and 4 corners.

9. Draw a figure with
0 sides and 0 corners.

Name _____

Ring two figures that have the same shape.
Write the name of the shape.

1. ⬚ ◯ ▢ ◁

$\underline{\text{square}}$

2. ◯ ▮ ▽ ◯

_____

3. ▮ ▭ ◯ ▷

_____

4. ◁ ◯ △ ▢

_____

Draw each figure.

5. triangle	6. circle	7. rectangle	8. square

9. Write how many of each figure.

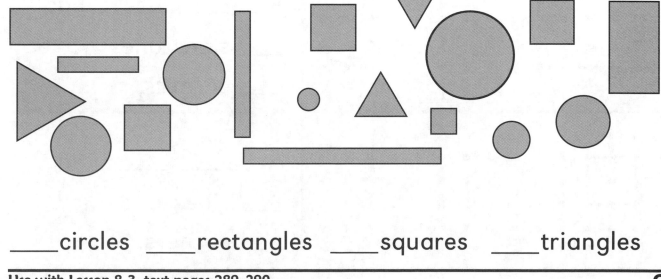

____circles   ____rectangles   ____squares   ____triangles

Match.

1.

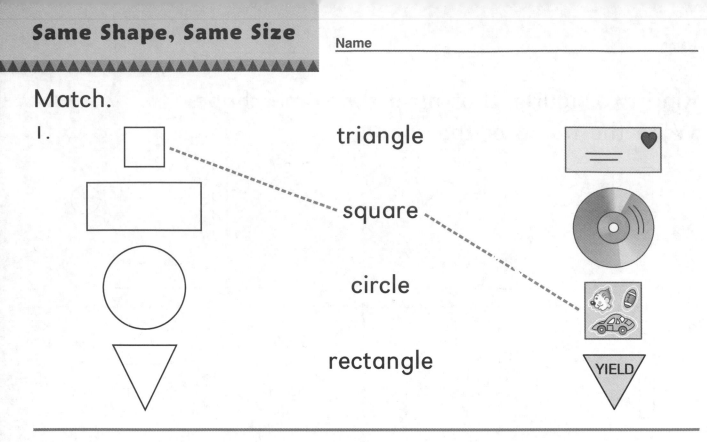

triangle

square

circle

rectangle

Draw a figure with the same size and the same shape. Name and describe each figure.

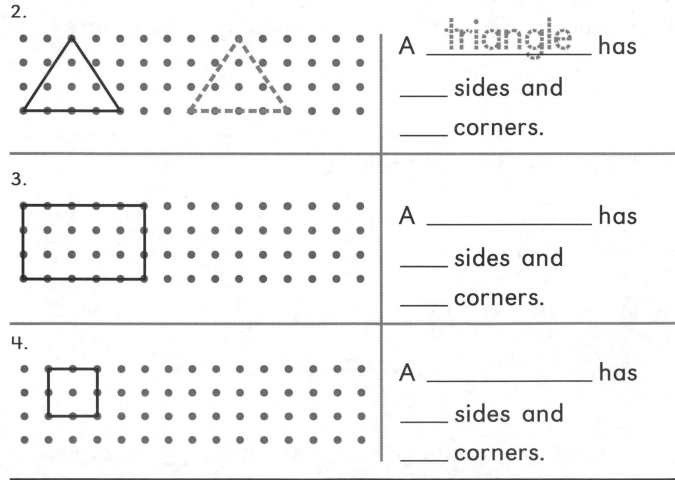

2. A ____triangle____ has

____ sides and

____ corners.

3. A _____ has

____ sides and

____ corners.

4. A _____ has

____ sides and

____ corners.

Use with Lessons 8-4 and 8-5, text pages 292–294.

Name _____

Ring the figure that shows matching parts.

1.

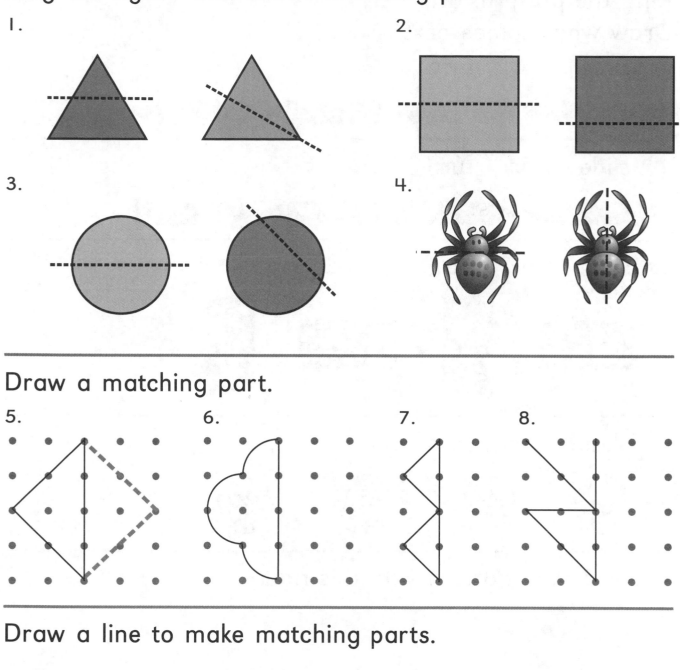

2.

3.

4.

Draw a matching part.

5.   6.   7.   8.

Draw a line to make matching parts.

9.   10.   11.   12.

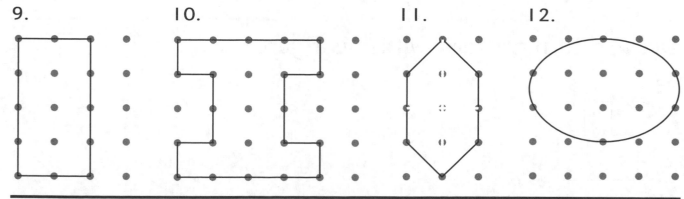

Ring the pattern.
Draw what comes next.

1. (slide)   or   turn

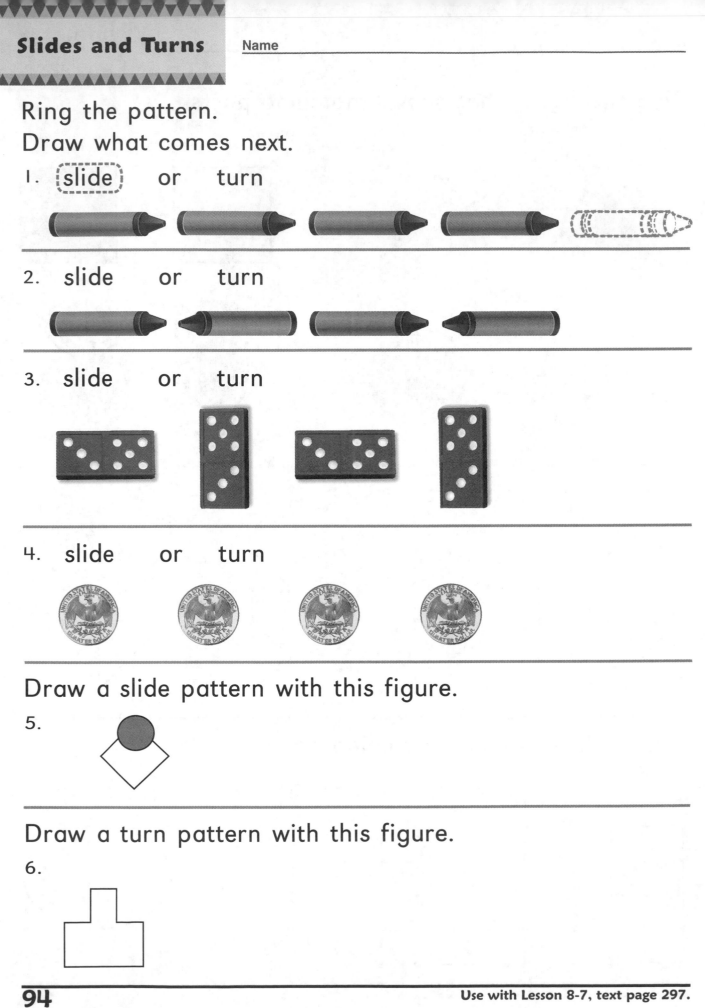

2. slide   or   turn

3. slide   or   turn

4. slide   or   turn

Draw a slide pattern with this figure.

5.

Draw a turn pattern with this figure.

6.

Use with Lesson 8-7, text page 297.

Name

✔ solids with curved surfaces.

1.

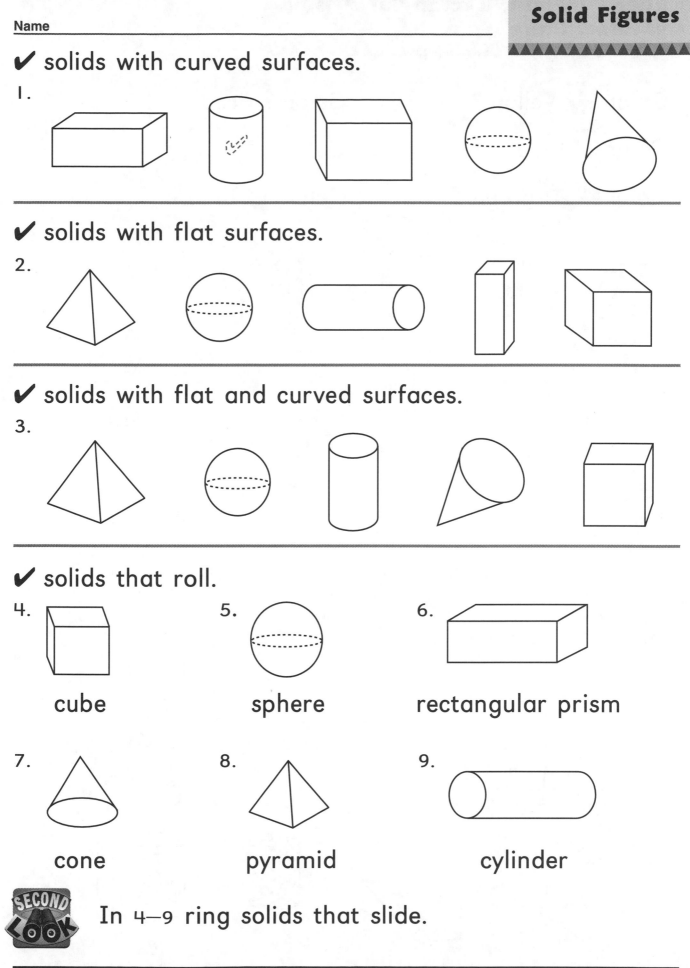

✔ solids with flat surfaces.

2.

✔ solids with flat and curved surfaces.

3.

✔ solids that roll.

4.   5.   6.

cube   sphere   rectangular prism

7.   8.   9.

cone   pyramid   cylinder

In 4–9 ring solids that slide.

Use with Lesson 8-8, text page 298.

Cube, Pyramid, Rectanglar Prism, Cylinder, Cone, Sphere

Name _____

Color 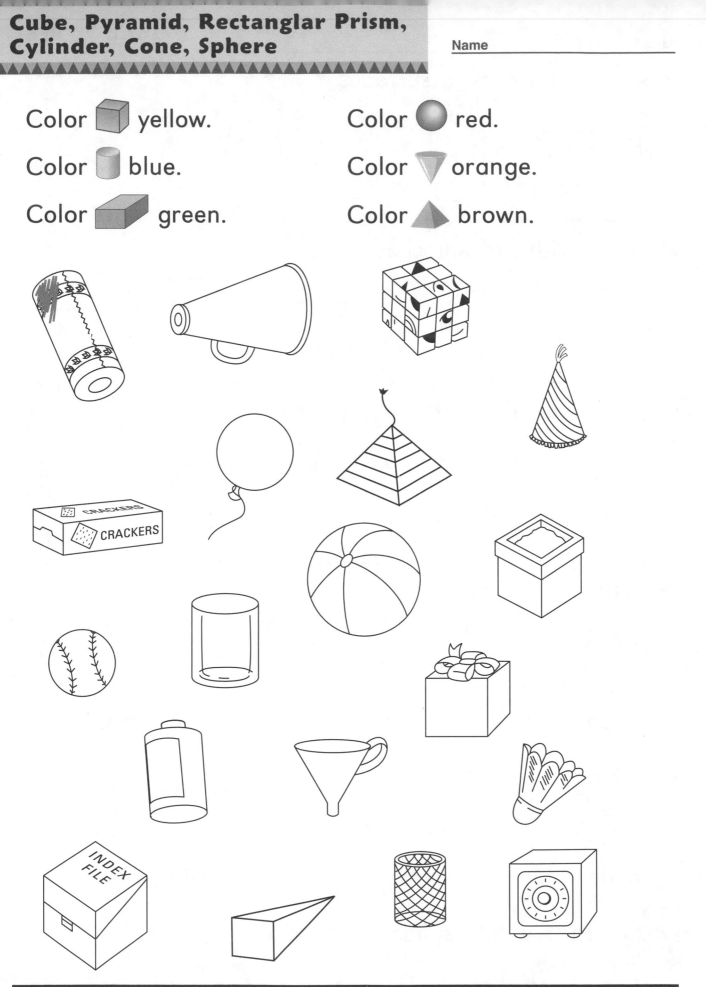 yellow.

Color blue.

Color green.

Color red.

Color orange.

Color brown.

Use with Lessons 8-9 and 8-10, text pages 299–302.

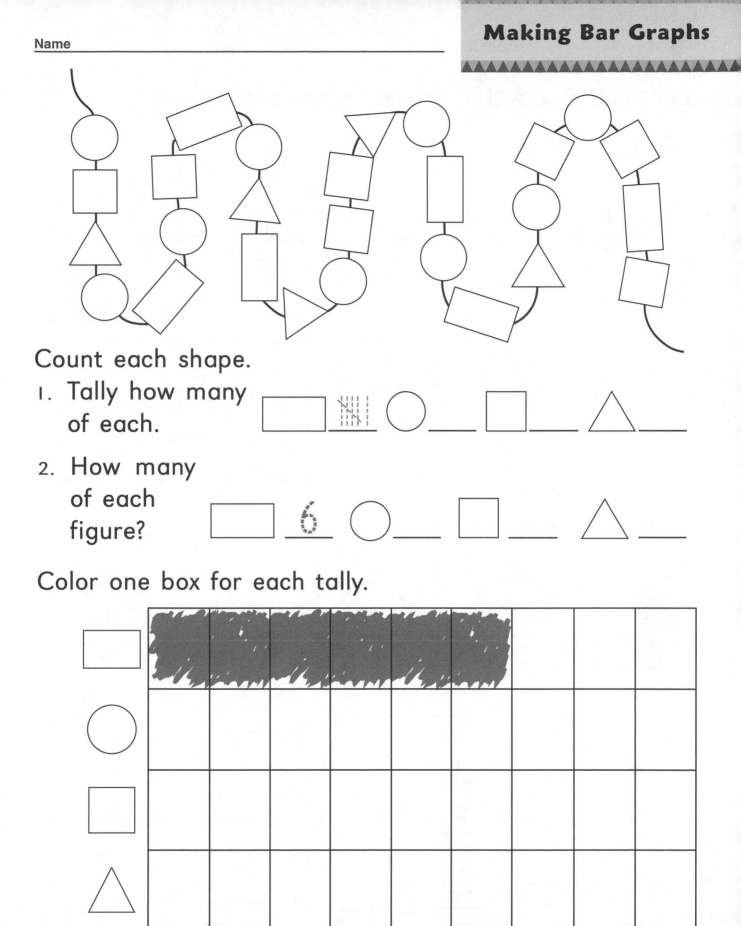

Count each shape.

1. Tally how many of each.

▭ |||| |||| ◯ ___   □ ___   △ ___

2. How many of each figure?

▭ _6_   ◯ ___   □ ___   △ ___

Color one box for each tally.

	0	1	2	3	4	5	6	7	8	9

Use with Lesson 8-11, text pages 303–304.

Ring each shape that shows equal parts.

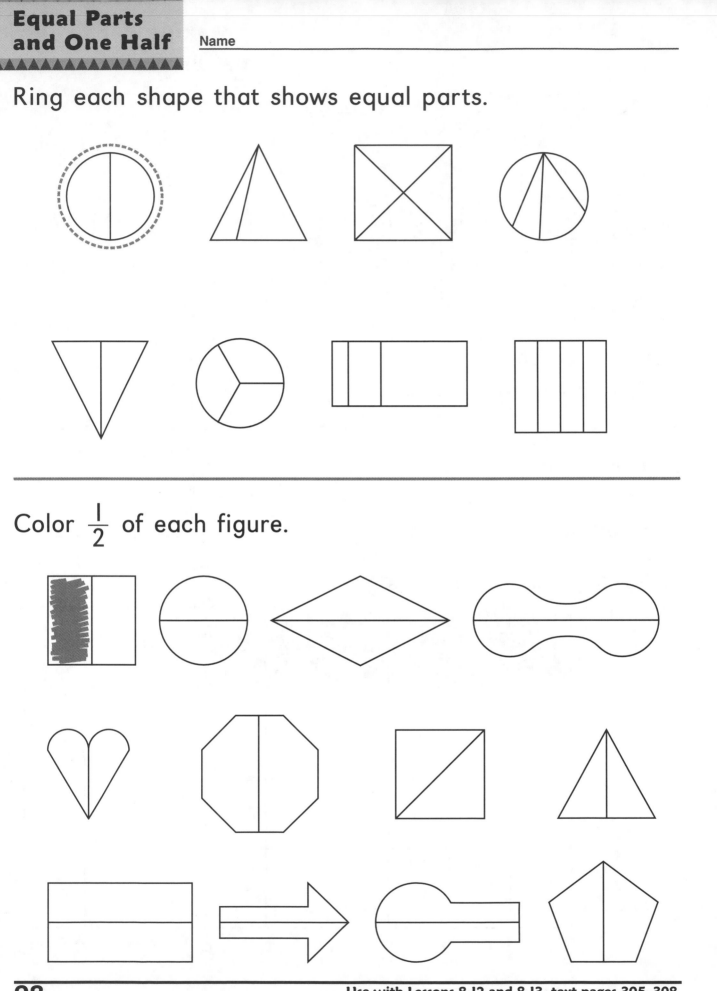

Color $\frac{1}{2}$ of each figure.

Use with Lessons 8-12 and 8-13, text pages 305–308.

Name _____

✔ each shape that shows 3 equal parts.

1.

✔ each shape that shows $\frac{1}{3}$ shaded.

2.

Color $\frac{1}{3}$ of each figure.

3.

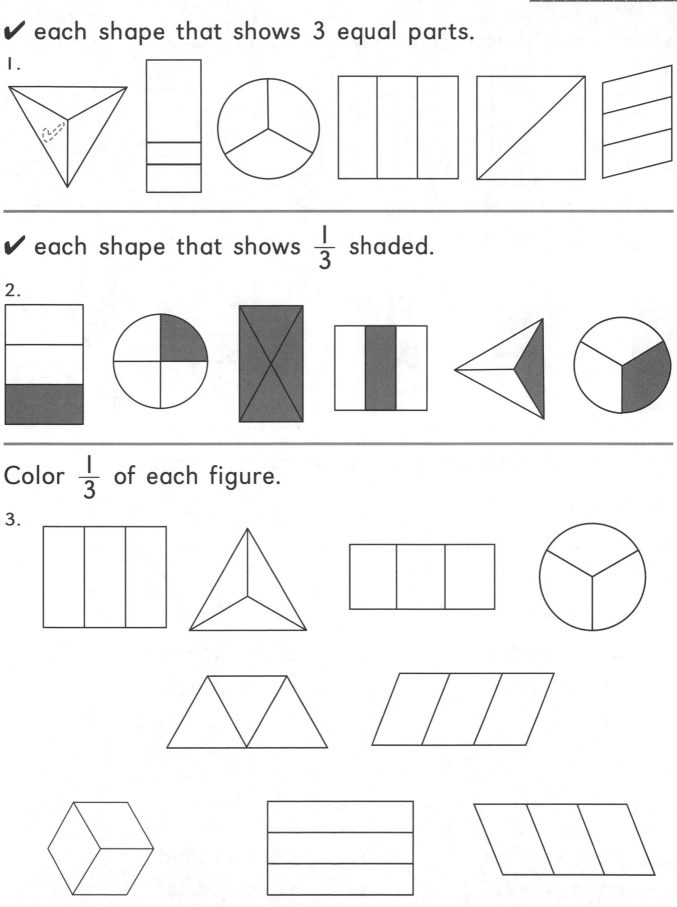

Color each shape that shows 4 equal parts.

1.

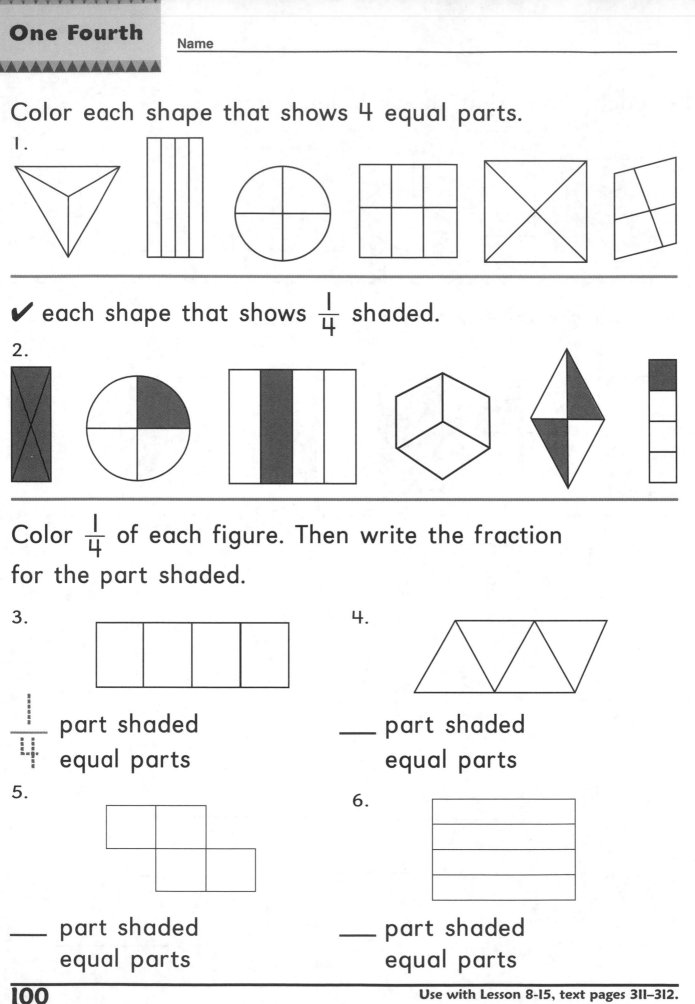

✔ each shape that shows $\frac{1}{4}$ shaded.

2.

Color $\frac{1}{4}$ of each figure. Then write the fraction for the part shaded.

3.

$\frac{1}{4}$ part shaded
equal parts

4.

___ part shaded
equal parts

5.

___ part shaded
equal parts

6.

___ part shaded
equal parts

Use with Lesson 8-15, text pages 311–312.

Name _____

What part of a set is missing?
Ring the fraction.

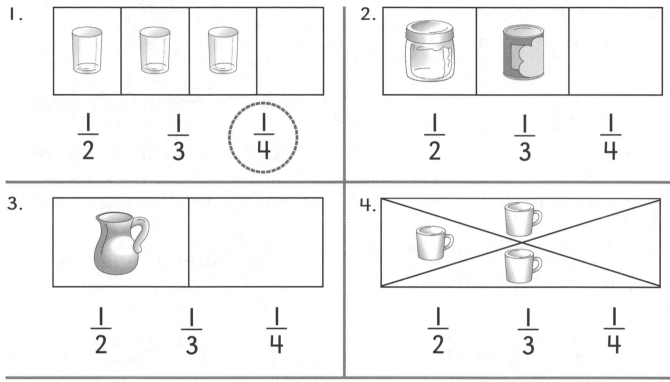

1.

$\frac{1}{2}$ $\frac{1}{3}$ $\boxed{\frac{1}{4}}$

2.

$\frac{1}{2}$ $\frac{1}{3}$ $\frac{1}{4}$

3.

$\frac{1}{2}$ $\frac{1}{3}$ $\frac{1}{4}$

4.

$\frac{1}{2}$ $\frac{1}{3}$ $\frac{1}{4}$

Color part of each set.

5. one half

6. one third

7. one fourth

Write the fraction that shows
what part is shaded.

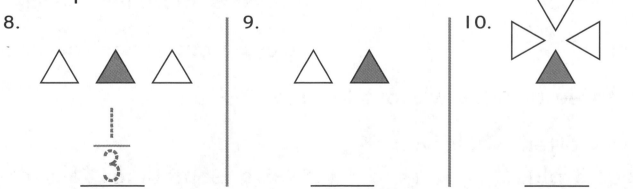

8.

$\frac{1}{3}$

9.

_____

10.

_____

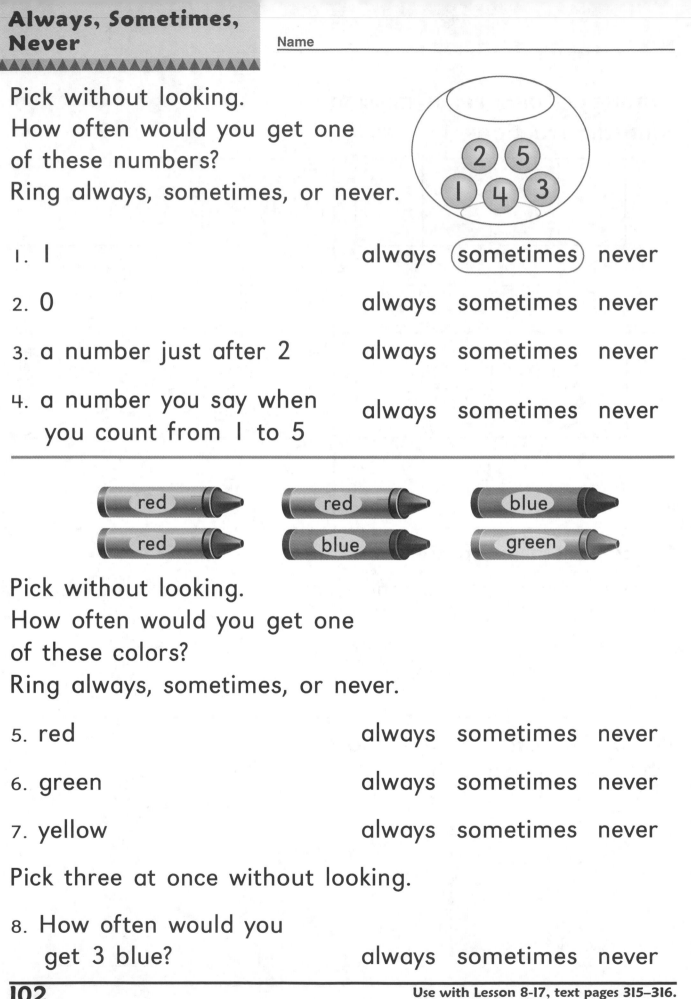

Pick without looking.
How often would you get one
of these numbers?
Ring always, sometimes, or never.

1. 1   always (sometimes) never

2. 0   always   sometimes   never

3. a number just after 2   always   sometimes   never

4. a number you say when
   you count from 1 to 5   always   sometimes   never

Pick without looking.
How often would you get one
of these colors?
Ring always, sometimes, or never.

5. red   always   sometimes   never

6. green   always   sometimes   never

7. yellow   always   sometimes   never

Pick three at once without looking.

8. How often would you
   get 3 blue?   always   sometimes   never

Use with Lesson 8-17, text pages 315–316.

Color to show the different ways
to make a badge.

1. You have a ▢ shaped badge and
   a ◯ shaped badge. You have
   a yellow ⭐ and a red ⭐.

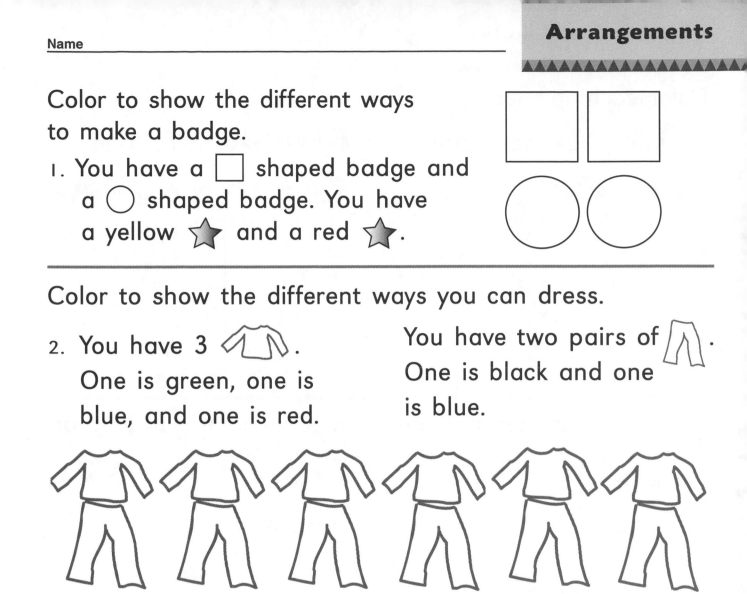

Color to show the different ways you can dress.

2. You have 3 👕.
   One is green, one is
   blue, and one is red.

   You have two pairs of 👖.
   One is black and one
   is blue.

3. You have I yellow, I red, and I blue bead.
   How many different ways can you order the 3 beads?
   Color to show the different ways.

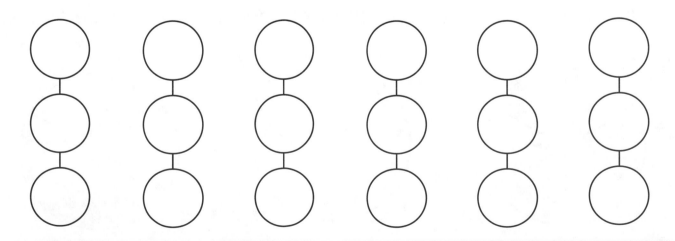

More or Less Likely

Name _____

Color each spinner to land on

1. 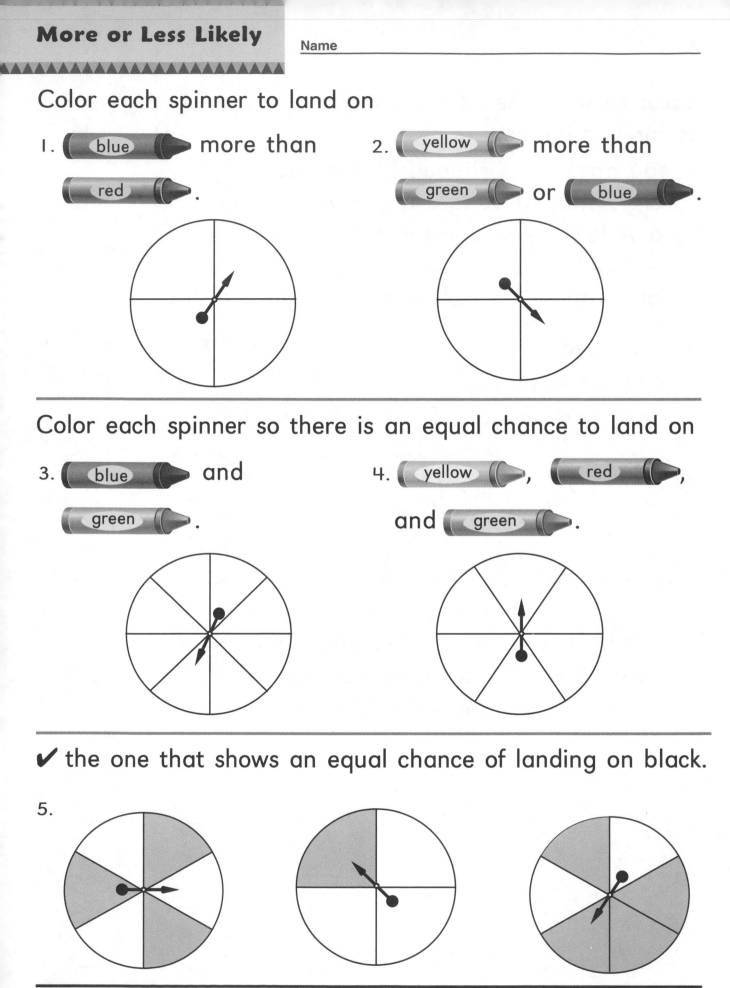 blue more than red.

2. yellow more than green or blue.

Color each spinner so there is an equal chance to land on

3. blue and green.

4. yellow, red, and green.

✔ the one that shows an equal chance of landing on black.

5.

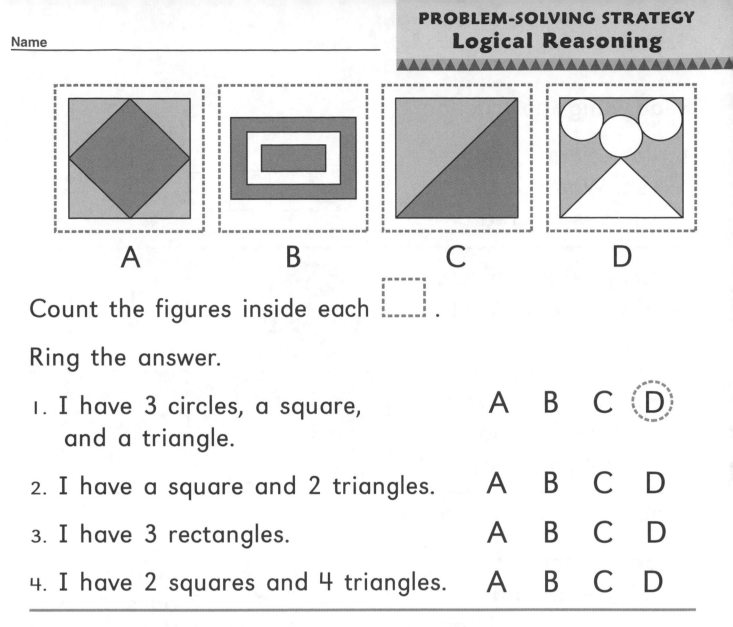

A   B   C   D

Count the figures inside each  .

Ring the answer.

1. I have 3 circles, a square, and a triangle.   A   B   C   (D)

2. I have a square and 2 triangles.   A   B   C   D

3. I have 3 rectangles.   A   B   C   D

4. I have 2 squares and 4 triangles.   A   B   C   D

Draw the part of the two space figures
that have the same shape.

5.

How many triangles inside each?

6. _____   7. _____

Read. Ring the fraction.

1. Pat colors 2 squares red. What part of the squares are red?

$\boxed{\dfrac{1}{2}}$ $\dfrac{1}{4}$ $\dfrac{1}{3}$

2. Karen eats 2 pieces of the pizza. What part is left?

$\dfrac{1}{2}$ $\dfrac{1}{4}$ $\dfrac{1}{3}$

3. Rick puts pennies in 3 sections. What part of the box is not full?

$\dfrac{1}{3}$ $\dfrac{1}{2}$ $\dfrac{1}{4}$

Read. Color. Then ring the answer.

4. Ben colors one third of the balls. How many balls are not colored?

1 ball
2 balls
3 balls

5. One half of Lara's balloons are blue. How many balloons are blue?

1 balloon
2 balloons
3 balloons

6. Pia gave one part of her pie away. She ate one fourth. How many pieces are left?

1 piece
2 pieces
3 pieces

Name _____

About how many are there? Ring the estimate.

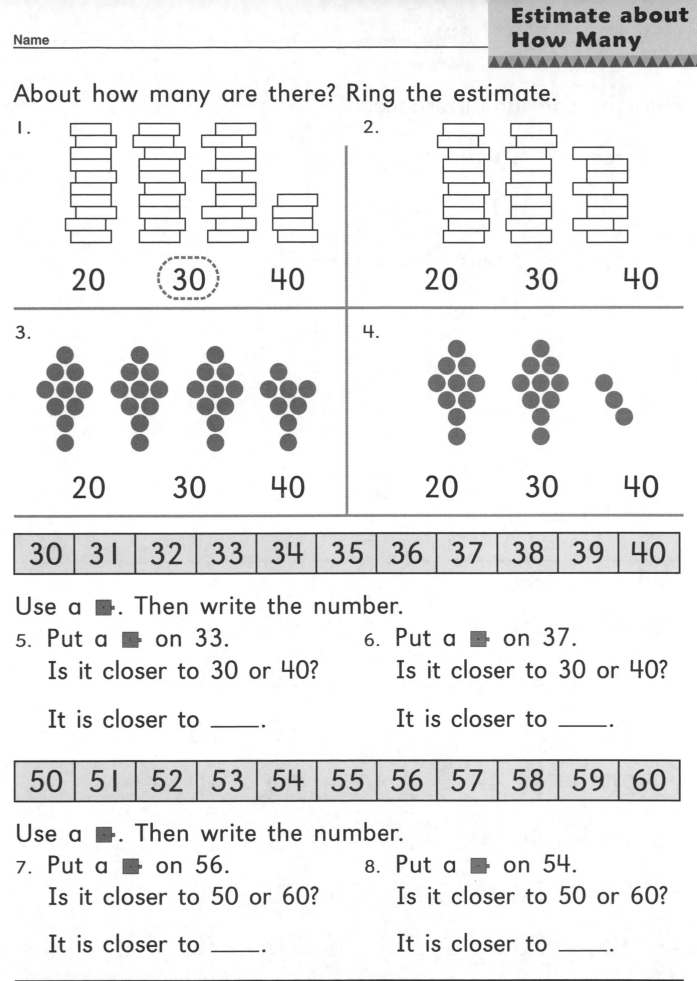

1. 20 (30) 40

2. 20 30 40

3. 20 30 40

4. 20 30 40

| 30 | 31 | 32 | 33 | 34 | 35 | 36 | 37 | 38 | 39 | 40 |

Use a ▪. Then write the number.

5. Put a ▪ on 33.
   Is it closer to 30 or 40?

   It is closer to _____.

6. Put a ▪ on 37.
   Is it closer to 30 or 40?

   It is closer to _____.

| 50 | 51 | 52 | 53 | 54 | 55 | 56 | 57 | 58 | 59 | 60 |

Use a ▪. Then write the number.

7. Put a ▪ on 56.
   Is it closer to 50 or 60?

   It is closer to _____.

8. Put a ▪ on 54.
   Is it closer to 50 or 60?

   It is closer to _____.

Add or Subtract Tens

Find the sum or difference.

1. 3 tens + 4 tens = ____ tens

   30 + 40 = ___

2. 8 tens − 3 tens = ____ tens

   80 − 30 = ___

3. 6 tens − 4 tens = ____ tens

   60 − 40 = ___

4. 7 tens + 2 tens = ____ tens

   70 + 20 = ___

Add or subtract.

5. 50 + 30 = ___   6. 80 − 30 = ___

7. 40 + 30 = ___   8. 70 − 30 = ___

9. 20 + 10 = ___   10. 30 − 10 = ___

11. 20 + 60 = ___   12. 80 − 60 = ___

13. 10 + 60 = ___   14. 70 − 10 = ___

15. 60 + 30 = ___   16. 90 − 30 = ___

17. 70 + 20 = ___   18. 90 − 20 = ___

Use with Lessons 9-2 and 9-3, text pages 337–338.

Name _____

Add and subtract.

1. 4 tens   40
  +2 tens   +20
  __6_ tens   ____

2. 6 dimes   60¢
  −2 dimes   −20¢
  __4_ dimes   ____ ¢

Find the sum or difference.
Draw or use 🪙 or ▭▭▭.

3.
   20   30   10   20¢   60¢
+50   +30   +70   +30¢   +30¢
__70_   ____   ____   ___¢   ___¢

4.
   90   70   90   80¢   70¢
−10   −10   −60   −40¢   −50¢
__80_   ____   ____   ___¢   ___¢

Find the missing numbers. Watch for + and −.

5.
   50
+ [10]
   60

   70
− 30
  [ ]

  [ ]
− 20
   70

   20
+ [ ]
   90

   10
+ 60
  [ ]

6.
   30
− [10]
   20

   80
− 30
  [ ]

  [ ]
− 30
   10

   20
+ 20
  [ ]

  [ ]
− 70
   20

Use with Lesson 9-4, text pages 339–340.

Add Tens and Ones

Use 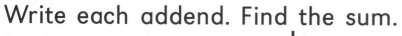 and ▪. Add.

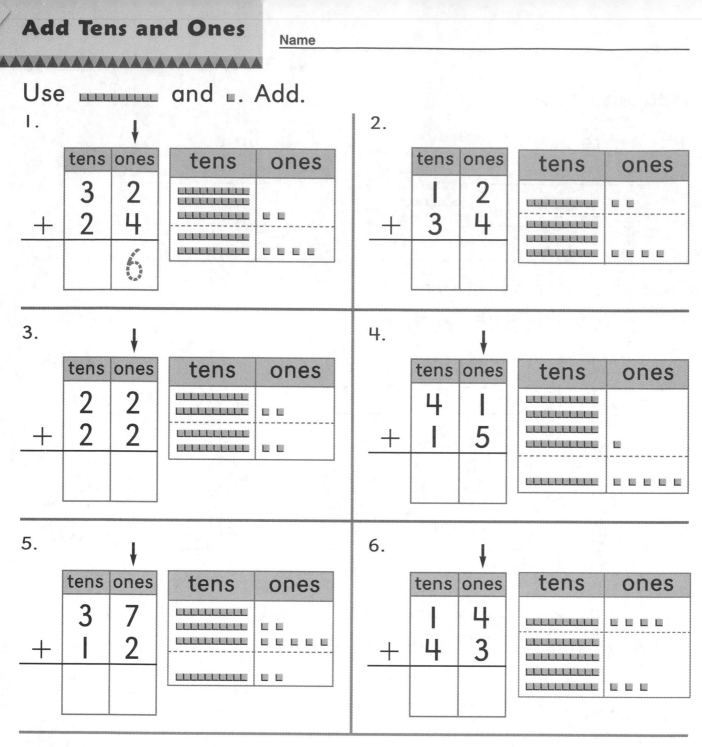

1.

tens	ones
3	2
+ 2	4
	6

2.

tens	ones
1	2
+ 3	4

3.

tens	ones
2	2
+ 2	2

4.

tens	ones
4	1
+ 1	5

5.

tens	ones
3	7
+ 1	2

6.

tens	ones
1	4
+ 4	3

Write each addend. Find the sum.

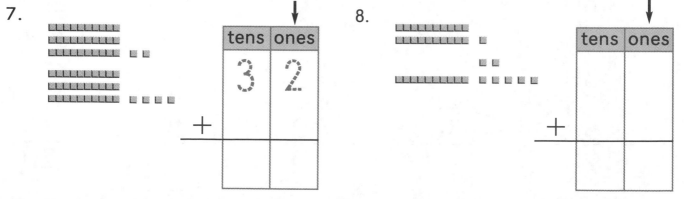

7.

tens	ones
3	2
+	

8.

tens	ones
+	

Use with Lesson 9-5, text pages 341–342.

Name _____

Add. Draw or use ▭▭▭▭ and ▪ to check.

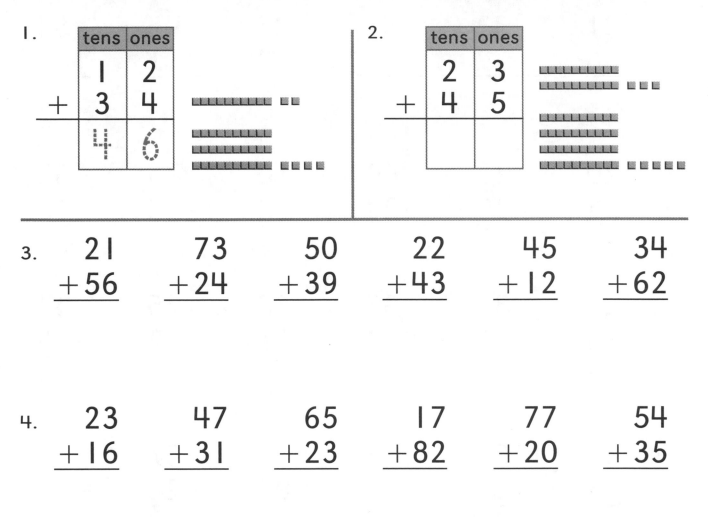

1.

tens	ones
1	2
+ 3	4
4	6

2.

tens	ones
2	3
+ 4	5

3.
$$21 + 56$$ $$73 + 24$$ $$50 + 39$$ $$22 + 43$$ $$45 + 12$$ $$34 + 62$$

4.
$$23 + 16$$ $$47 + 31$$ $$65 + 23$$ $$17 + 82$$ $$77 + 20$$ $$54 + 35$$

Find the sum. Change the order to check.

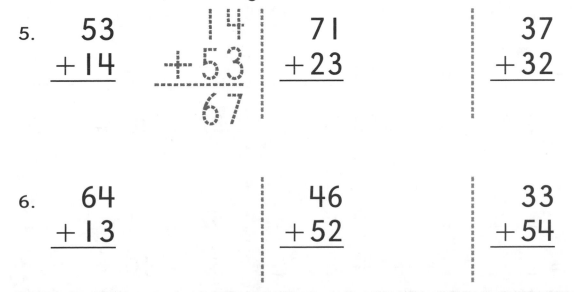

5.
$$53 + 14$$ $$14 + 53 \over 67$$ $$71 + 23$$ $$37 + 32$$

6.
$$64 + 13$$ $$46 + 52$$ $$33 + 54$$

Use with Lesson 9-6, text pages 343–344.

111

Add Tens or Ones

Name _____

Add mentally. Write the sum.

1.
$$21 + 6 = 27$$
$$33 + 2$$
$$51 + 3$$
$$27 + 2$$
$$45 + 1$$
$$54 + 2$$

2.
$$20 + 56$$
$$43 + 30$$
$$64 + 20$$
$$10 + 32$$
$$75 + 20$$
$$52 + 10$$

3.
$$20 + 6$$
$$32 + 40$$
$$35 + 30$$
$$60 + 11$$
$$25 + 2$$
$$44 + 5$$

4.
$$83 + 6$$
$$50 + 31$$
$$26 + 1$$
$$14 + 80$$
$$57 + 20$$
$$25 + 4$$

Add mentally. Watch for tens or ones.

5.
$$63 + 30 = 93 \qquad 41 + 5 = \underline{\quad} \qquad 37 + 2 = \underline{\quad}$$
$$63 + 3 = 66 \qquad 41 + 50 = \underline{\quad} \qquad 37 + 20 = \underline{\quad}$$

6.
$$16 + 2 = \underline{\quad} \qquad 24 + 40 = \underline{\quad} \qquad 52 + 1 = \underline{\quad}$$
$$16 + 20 = \underline{\quad} \qquad 24 + 4 = \underline{\quad} \qquad 52 + 10 = \underline{\quad}$$

112

Use with Lesson 9-7, text pages 345–346.

Name _____

Ring the ▭▭▭▭ and ▪ you subtract. Write the difference.

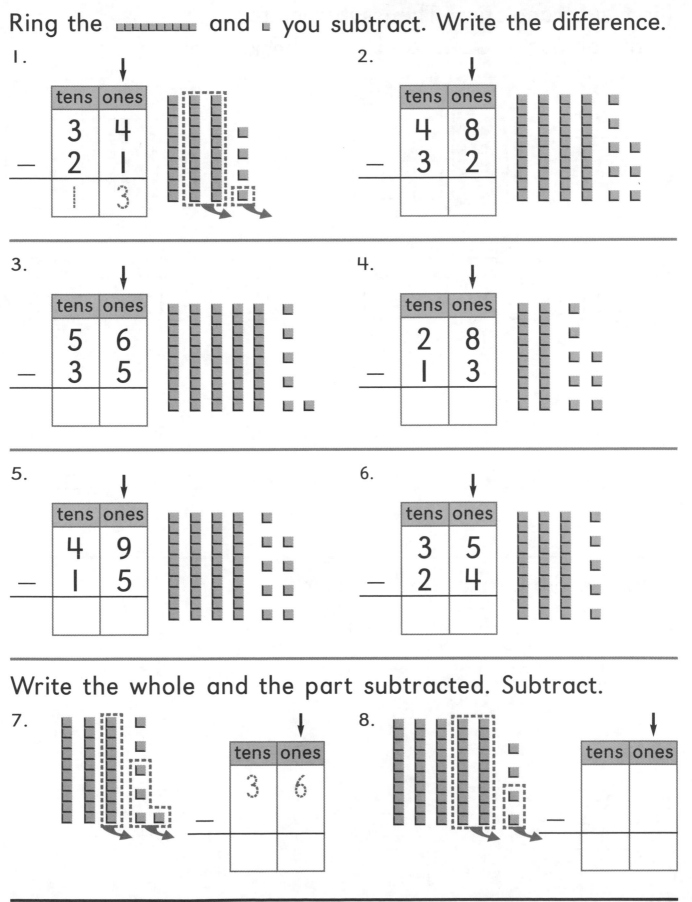

1.

tens	ones
3	4
− 2	1
	3

2.

tens	ones
4	8
− 3	2

3.

tens	ones
5	6
− 3	5

4.

tens	ones
2	8
− 1	3

5.

tens	ones
4	9
− 1	5

6.

tens	ones
3	5
− 2	4

Write the whole and the part subtracted. Subtract.

7.

tens	ones
3	6
−	

8.

tens	ones
−	

More Subtracting Tens and Ones

Name _____

First subtract ones. Then subtract tens.
Draw or use ▭▭▭ and ▪ to check.

1.
$$57 - 24 = 33$$

tens	ones

$$78 - 46$$

tens	ones

2.
$$56 - 21$$ $$77 - 24$$ $$53 - 31$$ $$43 - 22$$ $$85 - 12$$ $$68 - 25$$

3.
$$26 - 16$$ $$97 - 31$$ $$65 - 44$$ $$82 - 51$$ $$73 - 62$$ $$54 - 31$$

Find the difference. Add to check.

4.
$$54 - 13$$

$$41 + 13 = 54$$

$$79 - 23$$

$$97 - 32$$

5.
$$67 - 13$$

$$86 - 52$$

$$77 - 54$$

Use with Lesson 9-9, text pages 349–350.

Subtract mentally. Write the difference.

1.
49	23	56	27	25	98
− 6	− 2	− 3	− 2	− 1	− 3
43					

2.
56	93	68	77	45	39
−40	−30	−50	−40	−20	− 7
16					

3.
68	82	35	60	25	55
− 6	−40	− 3	−10	− 2	− 5

4.
88	57	26	19	67	37
− 6	−30	− 1	− 8	−20	− 5

Subtract mentally. Watch for tens and ones.

5.

$63 - 30 = 33$ $48 - 2 = \underline{\hphantom{00}}$ $37 - 1 = \underline{\hphantom{00}}$

$63 - 3 = 60$ $48 - 20 = \underline{\hphantom{00}}$ $37 - 10 = \underline{\hphantom{00}}$

6.

$76 - 5 = \underline{\hphantom{00}}$ $55 - 40 = \underline{\hphantom{00}}$ $89 - 6 = \underline{\hphantom{00}}$

$76 - 50 = \underline{\hphantom{00}}$ $55 - 4 = \underline{\hphantom{00}}$ $89 - 60 = \underline{\hphantom{00}}$

Estimating Sums and Differences; Add Money

Name _____

Estimate the answer. Use a ⟷.

```
←——+——+——+——+——+——+——+——+——+——+——+——→
  30 31 32 33 34 35 36 37 38 39 40
```

1. $32 + 37 = $ __?__

   $30 + 40 = $ _70_

2. $39 - 31 = $ __?__

   $40 - 30 = $ _10_

3. $28 + 21$ is about ____. $23 + 26$ is about ____.

4. $50 - 21$ is about ____. $50 - 26$ is about ____.

5. $43 + 41$ is about ____. $49 + 42$ is about ____.

6. $81 - 49$ is about ____. $92 - 44$ is about ____.

Find the sum.

7.
```
  26¢      30¢      75¢      45¢      28¢
 +23¢     +17¢     +11¢     +23¢     +51¢
```

8.
```
  68¢      82¢      35¢      60¢      25¢
 +30¢     +15¢     +43¢     +39¢     +62¢
```

9.
```
  37¢      50¢      72¢      13¢      27¢
 +11¢     +43¢     +16¢     +64¢     +42¢
```

Use with Lessons 9-11 and 9-12, text pages 354–356.

Name _____

Subtract. Use 🪙 and 🪙.

1.

dimes	pennies
4	5
− 2	1
2	4

45¢
−21¢
24¢

2.

dimes	pennies
3	8
− 1	6

38¢
−16¢

3.

46¢	37¢	68¢	75¢	58¢
−23¢	−10¢	−25¢	−11¢	−31¢

4.

66¢	57¢	98¢	25¢	76¢
−30¢	−23¢	−36¢	−14¢	−34¢

5.

97¢	38¢	48¢	95¢	67¢
−25¢	−24¢	−33¢	−70¢	−54¢

6.

89¢	45¢	49¢	72¢	59¢
−15¢	−23¢	−18¢	−61¢	−47¢

Name _____

Work from left to right. Ring the part you do first.
Add or subtract mentally.

1. (53 − 30) + 4 = _27_

    23 + 4

2. 55 + 40 − 3 = ___

3. 65 − 20 + 3 = ___

4. 81 − 50 + 5 = ___

Add or subtract. Watch for + and −.

5.	25 +34	33 −21	51 +36	17 +82	45 −33	94 −63
6.	56 −45	35 +33	98 −45	77 +22	45 +23	39 −17
7.	32 +26	87 −45	48 −34	96 −71	25 +72	55 +32

Add the ones. Regroup 10 ones as 1 ten when you can.

1. 2 tens 5 ones + 7 ones

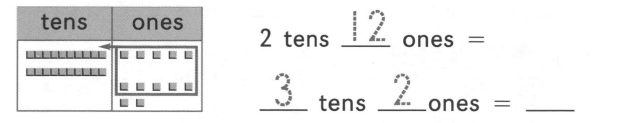

2 tens _12_ ones =

_3_ tens _2_ ones = ____

---

2. 3 tens 7 ones + 4 ones

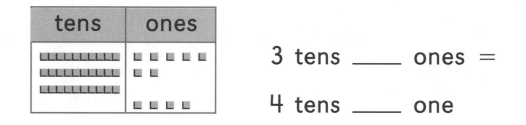

3 tens ____ ones =

4 tens ____ one

---

3. 4 tens 9 ones + 3 ones

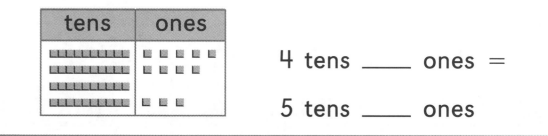

4 tens ____ ones =

5 tens ____ ones

Add ones. Regroup. Then add tens.

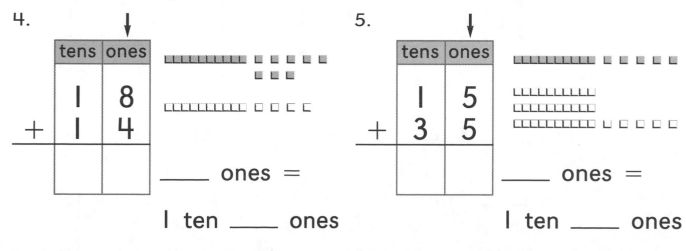

4.

tens	ones
1	8
+ 1	4

____ ones =

1 ten ____ ones

5.

tens	ones
1	5
+ 3	5

____ ones =

1 ten ____ ones

Regroup 1 ten as 10 ones when you need to.
Subtract after you regroup.

1. 3 tens 5 ones − 8 ones = __2__ tens __15__ ones − 8 ones

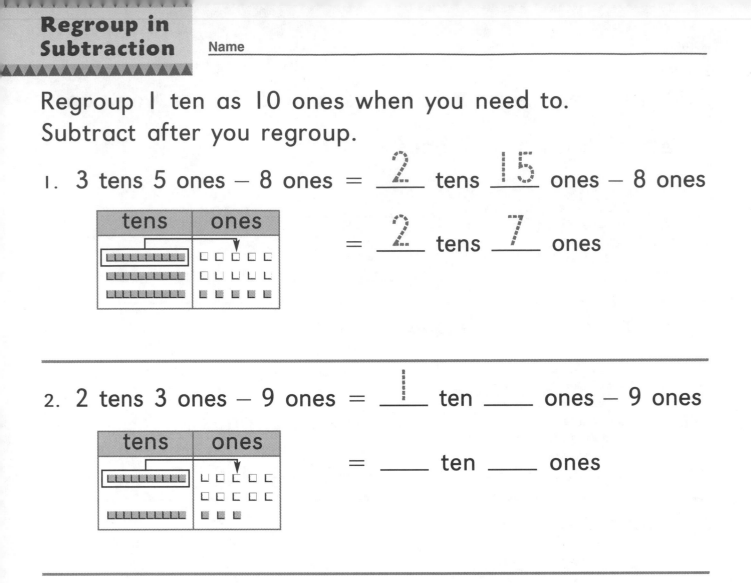

= __2__ tens __7__ ones

2. 2 tens 3 ones − 9 ones = __1__ ten ____ ones − 9 ones

= ____ ten ____ ones

Subtract. Regroup. Ring to take away.

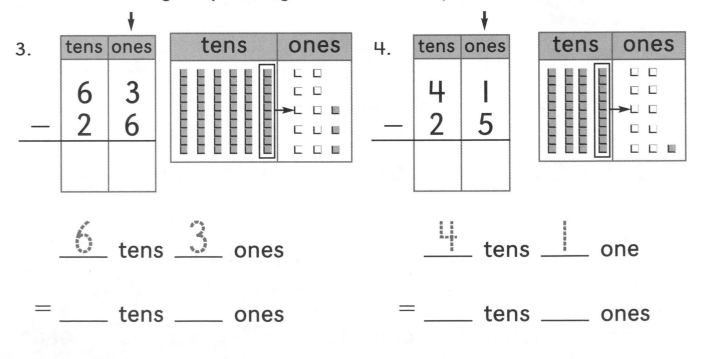

3.

tens	ones
6	3
− 2	6

__6__ tens __3__ ones

= ____ tens ____ ones

4.

tens	ones
4	1
− 2	5

__4__ tens __1__ one

= ____ tens ____ ones

Use with Lesson 9-16, text pages 363–364.

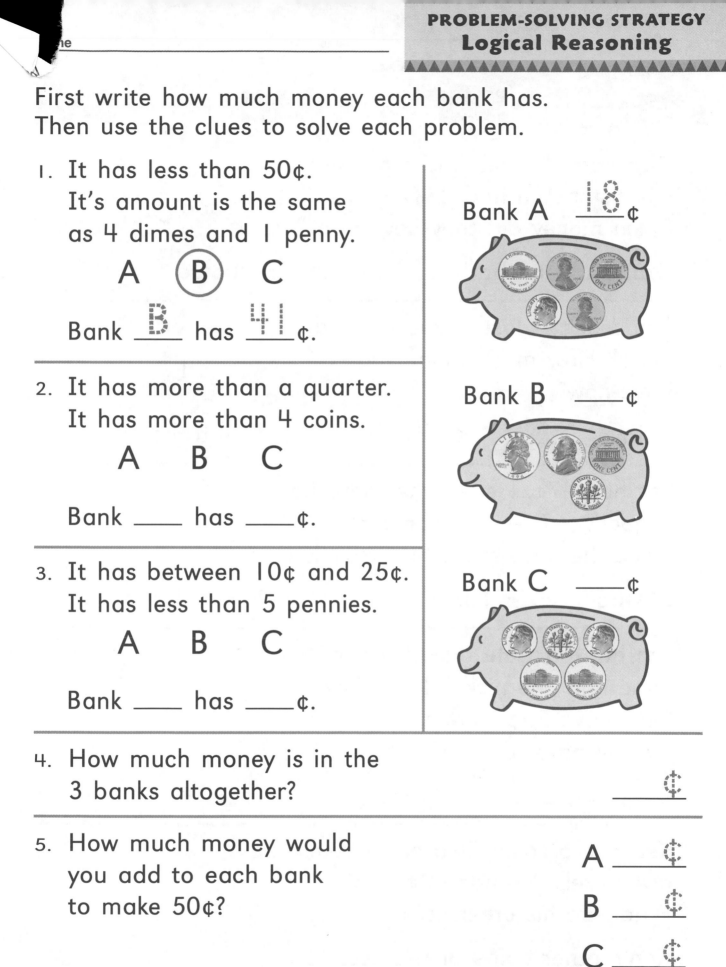

First write how much money each bank has.
Then use the clues to solve each problem.

1. It has less than 50¢.
   It's amount is the same
   as 4 dimes and 1 penny.

   A (B) C

   Bank __B__ has __41__¢.

Bank A __18__¢

2. It has more than a quarter.
   It has more than 4 coins.

   A   B   C

   Bank ___ has ___¢.

Bank B ___¢

3. It has between 10¢ and 25¢.
   It has less than 5 pennies.

   A   B   C

   Bank ___ has ___¢.

Bank C ___¢

4. How much money is in the
   3 banks altogether?   ___¢

5. How much money would
   you add to each bank
   to make 50¢?

   A ___¢
   B ___¢
   C ___¢

Read → Think → Write → Check

1. Dan had 2 dimes and 3 pennies. His sister Ann had 27¢. How much money did they have in all?

   Dan and Ann had _____ ¢ in all.

   (add)
   or
   subtract

2. I had 52¢. I lost a quarter and a nickel. How much money do I have now?

   I have _____ ¢.

   add
   or
   (subtract)

3. Pat had 47¢. Her mother gave her a quarter. Then she spent 35¢. Does she have more than 40¢ left?

   Pat has _____ ¢ left.

   Yes or No

4. Jim had 65¢. He gave Bill one dime and one nickel. Now they both have 50¢. How much money did Bill have to start?

   Bill had _____ ¢.

5. Ken has 5 coins. Two of his coins are nickels. He has 40¢ in all. What are his other coins?

   Ken's other coins are _____.

Name _____

Estimate about how many 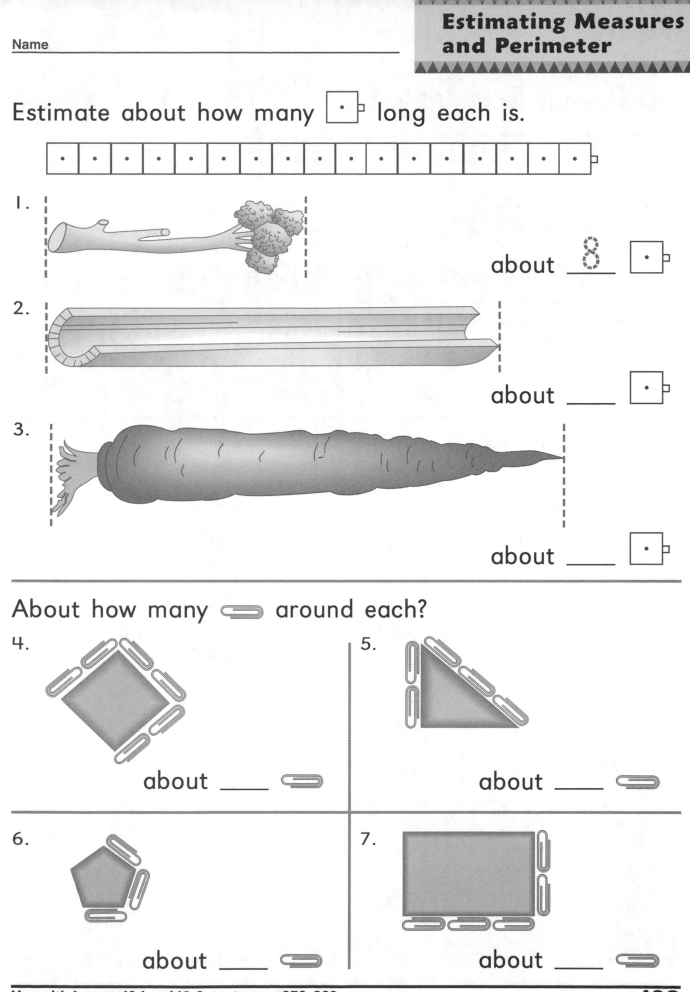 long each is.

1. about ___8___

2. about _____

3. about _____

About how many 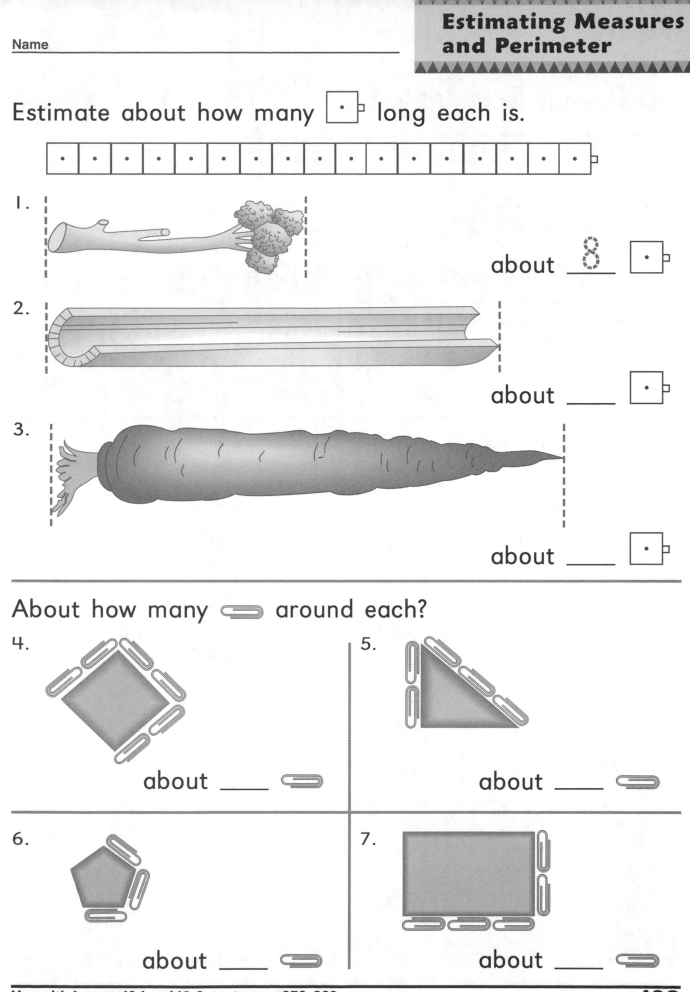 around each?

4. about _____

5. about _____

6. about _____

7. about _____

Measure the length in inches.

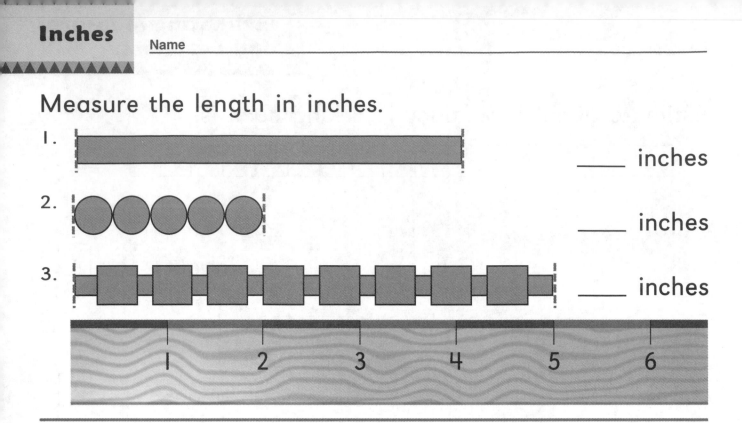

1. _____ inches

2. _____ inches

3. _____ inches

Measure the height in inches.

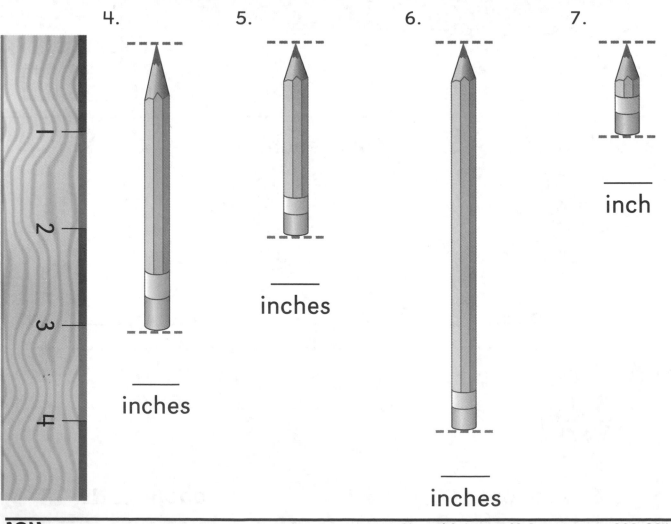

4. _____ inches

5. _____ inches

6. _____ inches

7. _____ inch

Use with Lesson 10-3, text pages 383–384.

Ring each real object that is more than 1 foot long.

1.

2.

3.

4. SEEDS

5.

6.

7.

8.

9.

SECOND LOOK ✔ each real object above that is less than 1 foot long.

Use with Lesson 10-4, text pages 385–386.

Ring the one that holds more.

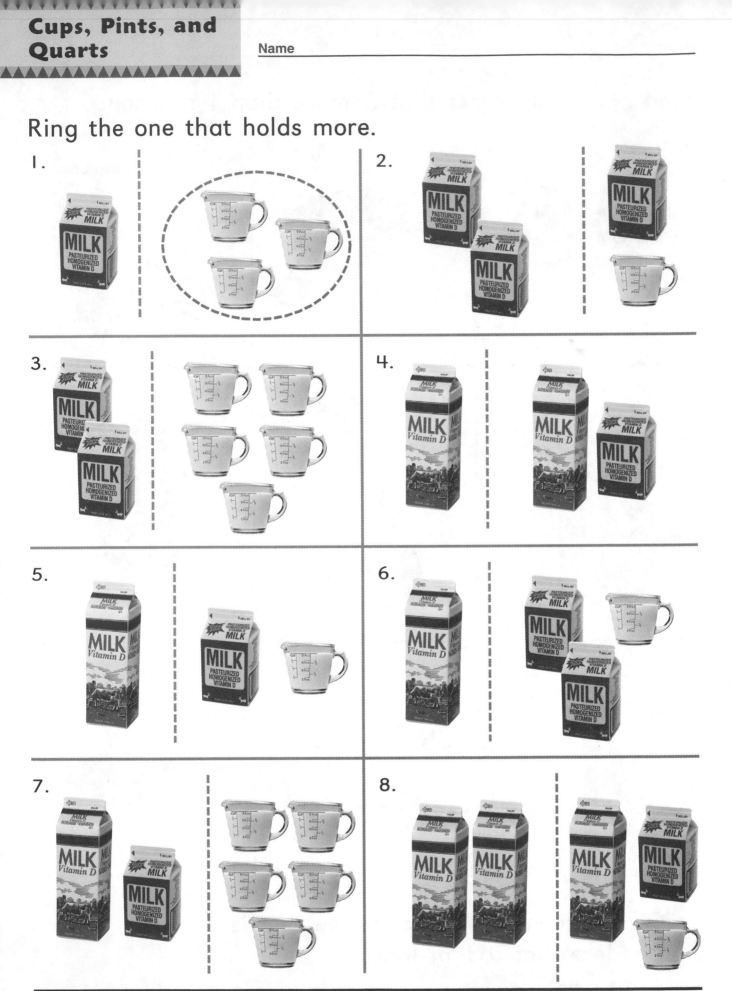

1.

2.

3.

4.

5.

6.

7.

8.

Use with Lessons 10-5 and 10-6, text pages 387–388.

Which weigh more than a pound? Ring.

1.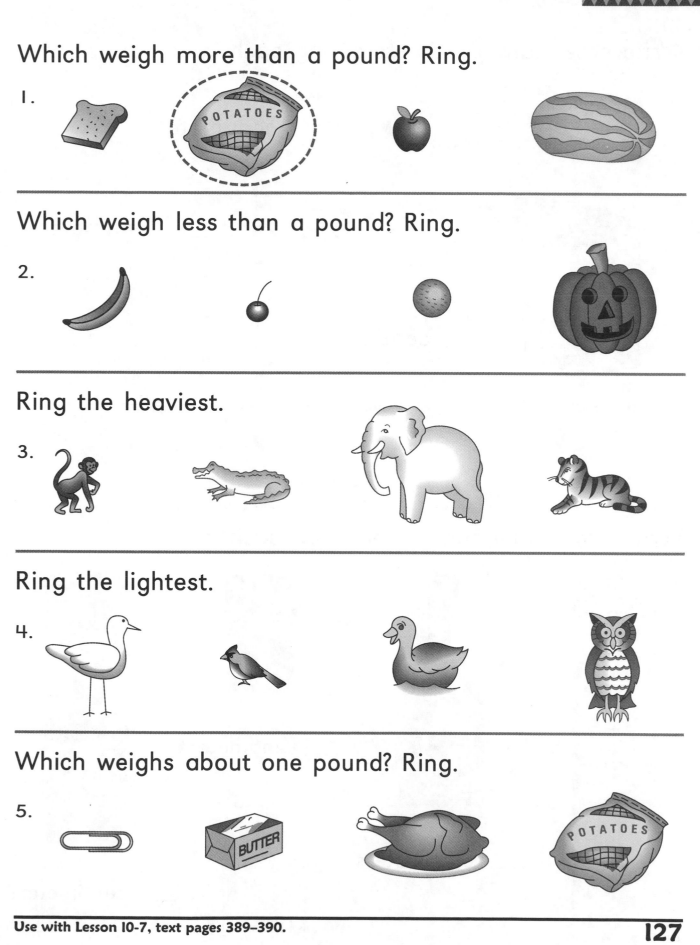

Which weigh less than a pound? Ring.

2.

Ring the heaviest.

3.

Ring the lightest.

4.

Which weighs about one pound? Ring.

5.

Name _____

Write how many centimeters long each is.

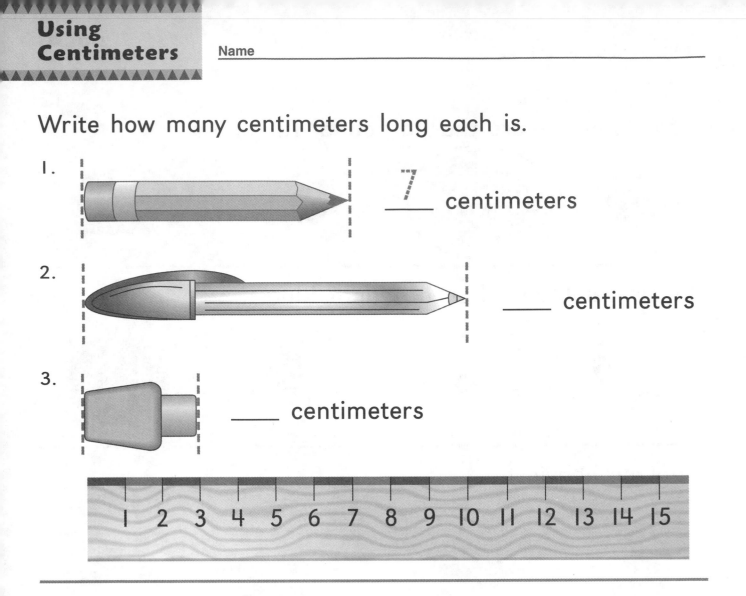

1. _7_ centimeters

2. ____ centimeters

3. ____ centimeters

| 1 | 2 | 3 | 4 | 5 | 6 | 7 | 8 | 9 | 10 | 11 | 12 | 13 | 14 | 15 |

Write how many centimeters high each is.

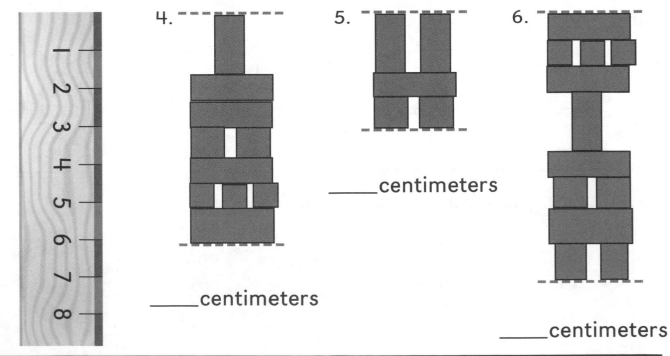

4. ____ centimeters

5. ____ centimeters

6. ____ centimeters

Use with Lessons 10-8 and 10-9, text pages 391–393.

Name _____

Ring each object that holds more than 1 liter.

1.

2.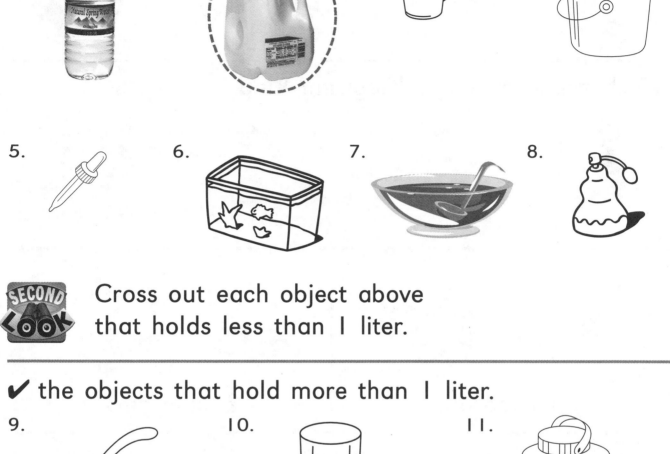

3.

4.

5.

6.

7.

8.

 SECOND LOOK Cross out each object above that holds less than 1 liter.

✔ the objects that hold more than 1 liter.

9.

10.

11.

✔ the objects that hold less than 1 liter.

12.

JUICE

13.

14.

Which are more than 1 kilogram? Ring.

1.

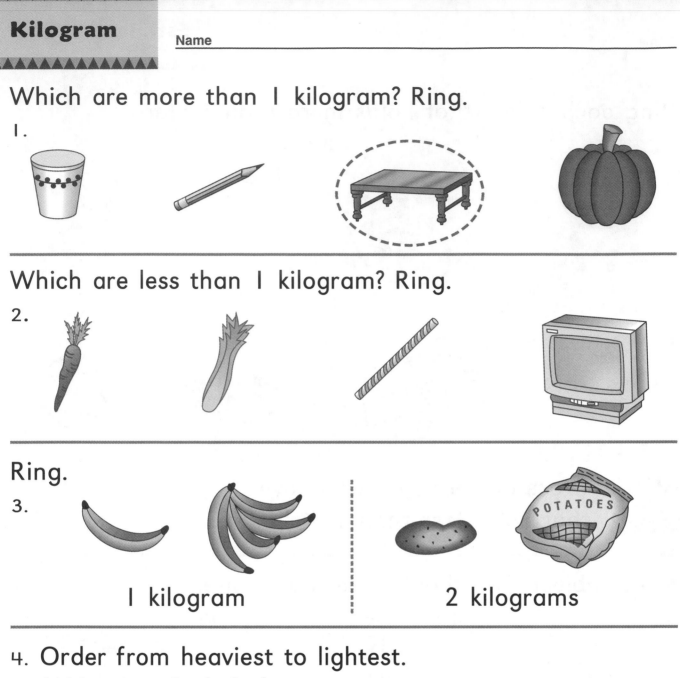

Which are less than 1 kilogram? Ring.

2.

Ring.

3.

1 kilogram   2 kilograms

4. Order from heaviest to lightest.
   Write 1st, 2nd, 3rd.

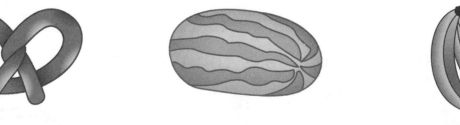

_____ _____ _____

 In 4 ring the object that is
about 1 kilogram.

Name _____

Which tool would you use to measure each?

Ring C for 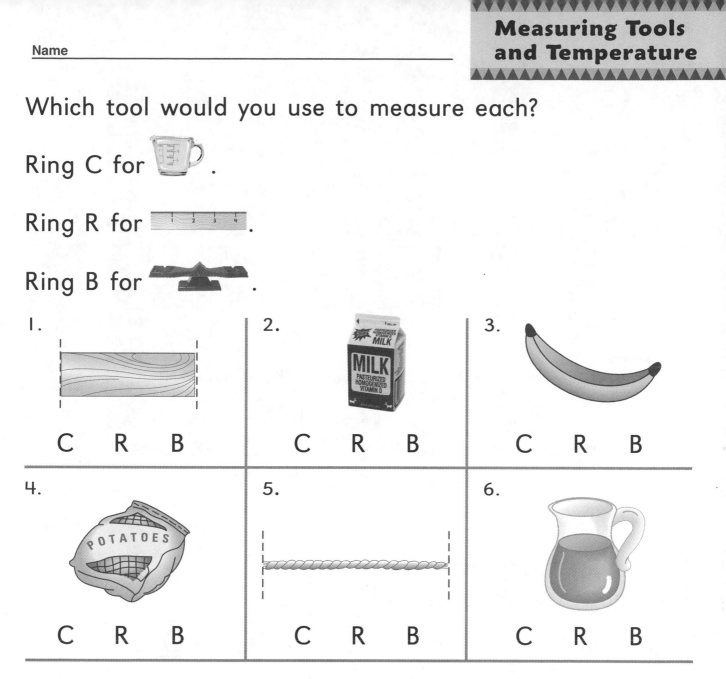 .

Ring R for .

Ring B for .

1.

C R B

2. MILK PASTEURIZED HOMOGENIZED VITAMIN D

C R B

3.

C R B

4. POTATOES

C R B

5.

C R B

6.

C R B

Ring the temperature. Color the thermometer.

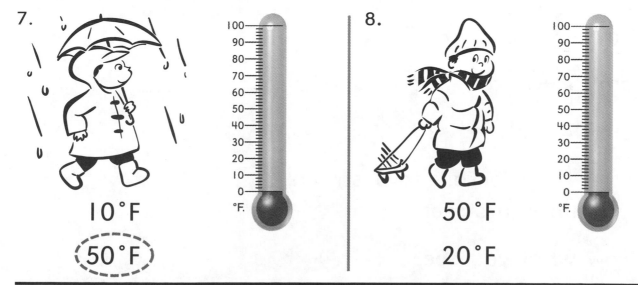

7.

10°F

(50°F)

8.

50°F

20°F

PROBLEM-SOLVING STRATEGY
Use a Map

Name _____

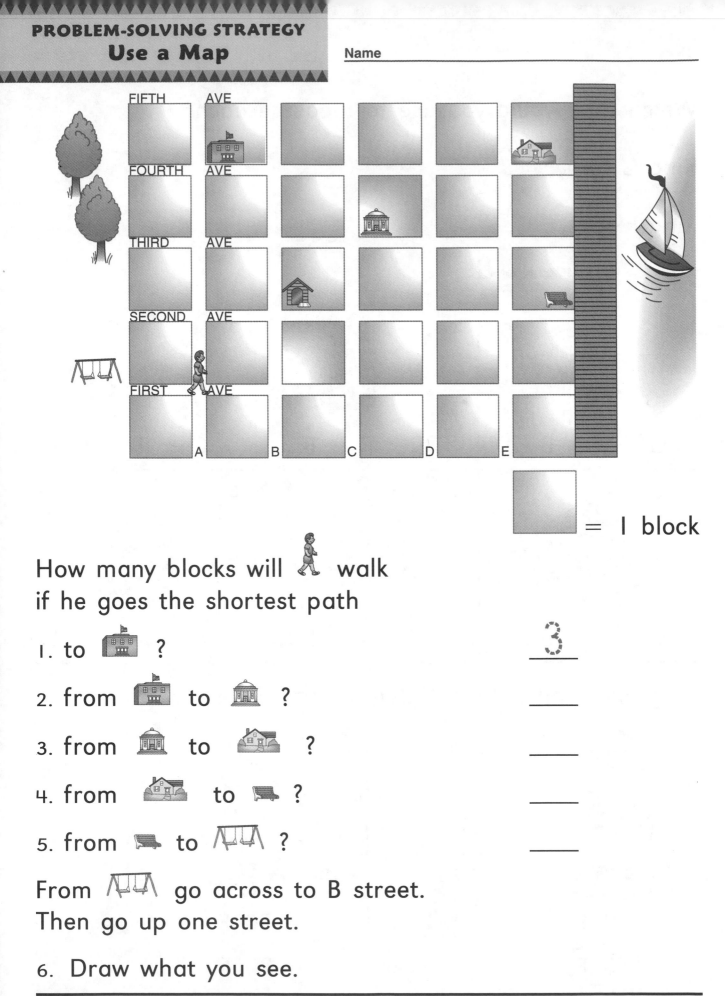

= 1 block

How many blocks will 🚶 walk
if he goes the shortest path

1. to 🏫 ?   _3_

2. from 🏫 to 🏛 ?   ____

3. from 🏛 to 🏠 ?   ____

4. from 🏠 to 🪑 ?   ____

5. from 🪑 to ⛓ ?   ____

From ⛓ go across to B street.
Then go up one street.

6. Draw what you see.

          Use with Lesson 10-15, text pages 403–404.

Name _____

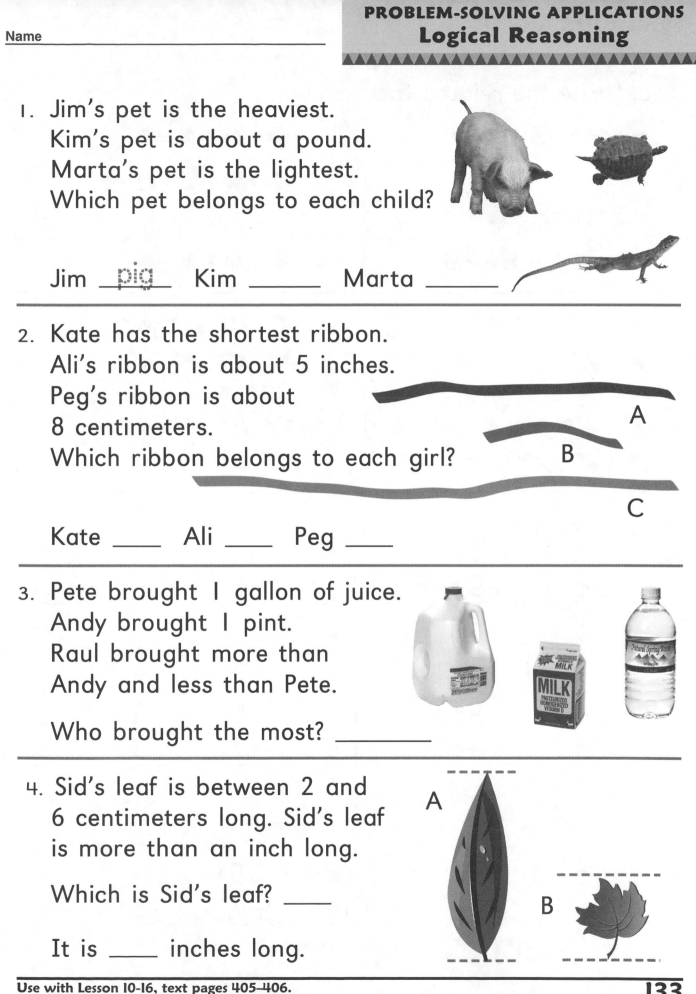

1. Jim's pet is the heaviest.
   Kim's pet is about a pound.
   Marta's pet is the lightest.
   Which pet belongs to each child?

   Jim __pig__   Kim _____   Marta _____

2. Kate has the shortest ribbon.
   Ali's ribbon is about 5 inches.
   Peg's ribbon is about
   8 centimeters.
   Which ribbon belongs to each girl?

   A

   B

   C

   Kate ____   Ali ____   Peg ____

3. Pete brought 1 gallon of juice.
   Andy brought 1 pint.
   Raul brought more than
   Andy and less than Pete.

   Who brought the most? _____

4. Sid's leaf is between 2 and
   6 centimeters long. Sid's leaf
   is more than an inch long.

   A

   Which is Sid's leaf? ____

   B

   It is ____ inches long.

Add. Write the related fact.

1.

$$\begin{array}{r} 4 \\ + 9 \\ \hline 13 \end{array}$$

$$\begin{array}{r} 9 \\ + 4 \\ \hline \end{array}$$

2.

$$\begin{array}{r} 6 \\ + 8 \\ \hline \end{array}$$

$$\begin{array}{r} \\ + \\ \hline \end{array}$$

Find the sum.

3.
$$\begin{array}{r} 6¢ \\ +6¢ \\ \hline 12¢ \end{array}$$
$$\begin{array}{r} 2¢ \\ +9¢ \\ \hline ¢ \end{array}$$
$$\begin{array}{r} 5¢ \\ +9¢ \\ \hline ¢ \end{array}$$
$$\begin{array}{r} 4 \\ +9 \\ \hline \end{array}$$
$$\begin{array}{r} 7 \\ +5 \\ \hline \end{array}$$
$$\begin{array}{r} 8 \\ +5 \\ \hline \end{array}$$

4.
$$\begin{array}{r} 7¢ \\ +7¢ \\ \hline ¢ \end{array}$$
$$\begin{array}{r} 9¢ \\ +4¢ \\ \hline ¢ \end{array}$$
$$\begin{array}{r} 8¢ \\ +6¢ \\ \hline ¢ \end{array}$$
$$\begin{array}{r} 6 \\ +7 \\ \hline \end{array}$$
$$\begin{array}{r} 6 \\ +5 \\ \hline \end{array}$$
$$\begin{array}{r} 3 \\ +8 \\ \hline \end{array}$$

5.
$$\begin{array}{r} 5¢ \\ +8¢ \\ \hline ¢ \end{array}$$
$$\begin{array}{r} 9¢ \\ +5¢ \\ \hline ¢ \end{array}$$
$$\begin{array}{r} 7¢ \\ +6¢ \\ \hline ¢ \end{array}$$
$$\begin{array}{r} 5 \\ +7 \\ \hline \end{array}$$
$$\begin{array}{r} 6 \\ +8 \\ \hline \end{array}$$
$$\begin{array}{r} 7 \\ +4 \\ \hline \end{array}$$

Use with Lesson 11-1, text pages 415–416.

Name _____

Subtract. Write the related fact.

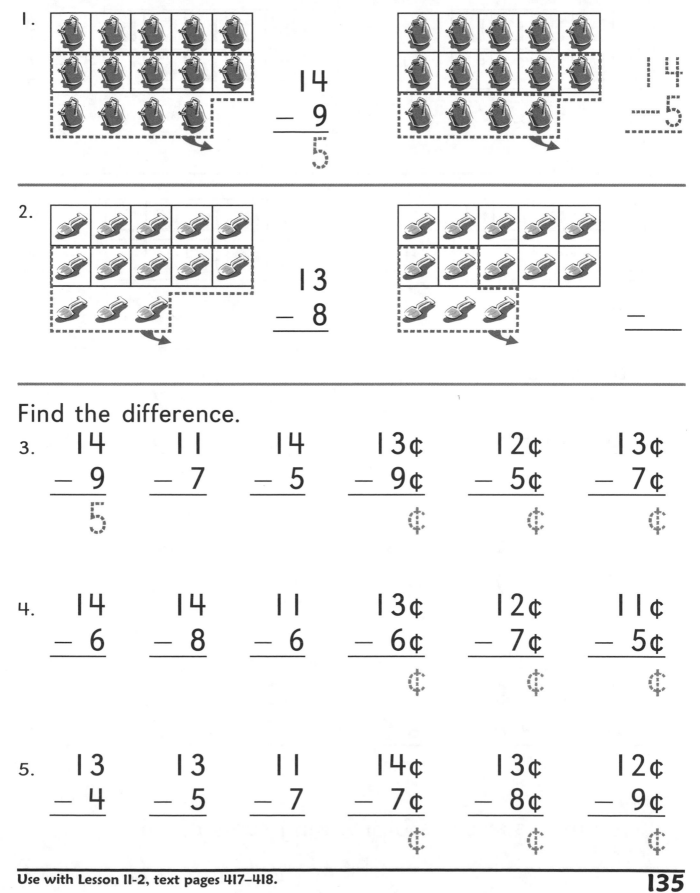

1.

$$\begin{array}{r} 14 \\ -\ 9 \\ \hline 5 \end{array}$$

$$\begin{array}{r} 14 \\ -5 \\ \hline \end{array}$$

2.

$$\begin{array}{r} 13 \\ -\ 8 \\ \hline \end{array}$$

$$\begin{array}{r} \_\_ \\ \hline \end{array}$$

Find the difference.

3.
$$\begin{array}{r} 14 \\ -\ 9 \\ \hline 5 \end{array}$$
$$\begin{array}{r} 11 \\ -\ 7 \\ \hline \end{array}$$
$$\begin{array}{r} 14 \\ -\ 5 \\ \hline \end{array}$$
$$\begin{array}{r} 13¢ \\ -\ 9¢ \\ \hline ¢ \end{array}$$
$$\begin{array}{r} 12¢ \\ -\ 5¢ \\ \hline ¢ \end{array}$$
$$\begin{array}{r} 13¢ \\ -\ 7¢ \\ \hline ¢ \end{array}$$

4.
$$\begin{array}{r} 14 \\ -\ 6 \\ \hline \end{array}$$
$$\begin{array}{r} 14 \\ -\ 8 \\ \hline \end{array}$$
$$\begin{array}{r} 11 \\ -\ 6 \\ \hline \end{array}$$
$$\begin{array}{r} 13¢ \\ -\ 6¢ \\ \hline ¢ \end{array}$$
$$\begin{array}{r} 12¢ \\ -\ 7¢ \\ \hline ¢ \end{array}$$
$$\begin{array}{r} 11¢ \\ -\ 5¢ \\ \hline ¢ \end{array}$$

5.
$$\begin{array}{r} 13 \\ -\ 4 \\ \hline \end{array}$$
$$\begin{array}{r} 13 \\ -\ 5 \\ \hline \end{array}$$
$$\begin{array}{r} 11 \\ -\ 7 \\ \hline \end{array}$$
$$\begin{array}{r} 14¢ \\ -\ 7¢ \\ \hline ¢ \end{array}$$
$$\begin{array}{r} 13¢ \\ -\ 8¢ \\ \hline ¢ \end{array}$$
$$\begin{array}{r} 12¢ \\ -\ 9¢ \\ \hline ¢ \end{array}$$

Use with Lesson 11-2, text pages 417–418.

Name _____

Draw the missing addend. Find the sum.

1.

$$\begin{array}{r} 7 \\ + 9 \\ \hline 16 \end{array}$$

$$\begin{array}{r} 9 \\ + 7 \\ \hline \end{array}$$

2.

$$\begin{array}{r} 7 \\ + 8 \\ \hline \end{array}$$

$$\begin{array}{r} 8 \\ + 7 \\ \hline \end{array}$$

Find the sum.

3.
$$\begin{array}{r} 8 \\ + 8 \\ \hline 16 \end{array}$$
$$\begin{array}{r} 5 \\ + 9 \\ \hline \end{array}$$
$$\begin{array}{r} 6 \\ + 9 \\ \hline \end{array}$$
$$\begin{array}{r} 4 \\ + 8 \\ \hline \end{array}$$
$$\begin{array}{r} 7 \\ + 8 \\ \hline \end{array}$$
$$\begin{array}{r} 8 \\ + 5 \\ \hline \end{array}$$

4.
$$\begin{array}{r} 7 \\ + 7 \\ \hline \end{array}$$
$$\begin{array}{r} 9 \\ + 7 \\ \hline \end{array}$$
$$\begin{array}{r} 8 \\ + 6 \\ \hline \end{array}$$
$$\begin{array}{r} 6 \\ + 7 \\ \hline \end{array}$$
$$\begin{array}{r} 6 \\ + 5 \\ \hline \end{array}$$
$$\begin{array}{r} 3 \\ + 8 \\ \hline \end{array}$$

5.
$$\begin{array}{r} 5 \\ + 8 \\ \hline \end{array}$$
$$\begin{array}{r} 9 \\ + 6 \\ \hline \end{array}$$
$$\begin{array}{r} 7 \\ + 6 \\ \hline \end{array}$$
$$\begin{array}{r} 5 \\ + 7 \\ \hline \end{array}$$
$$\begin{array}{r} 8 \\ + 7 \\ \hline \end{array}$$
$$\begin{array}{r} 7 \\ + 9 \\ \hline \end{array}$$

SECOND LOOK In 3–5 ✔ sums of 15. Ring sums of 16.

Name _____

Subtract. Write the related fact.

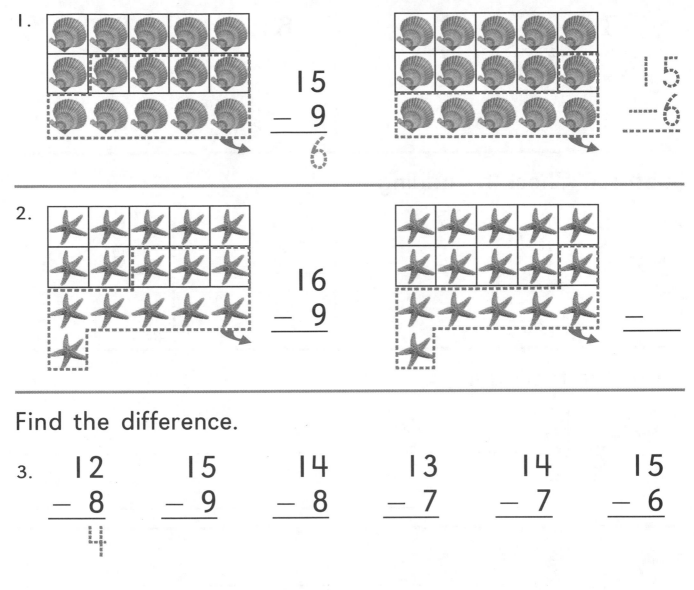

1.

$$\begin{array}{r} 15 \\ -\ 9 \\ \hline 6 \end{array}$$

$$\begin{array}{r} 15 \\ -\ 6 \\ \hline \end{array}$$

2.

$$\begin{array}{r} 16 \\ -\ 9 \\ \hline \end{array}$$

$$\begin{array}{r} - \\ \hline \end{array}$$

Find the difference.

3.
$$\begin{array}{r} 12 \\ -\ 8 \\ \hline 4 \end{array}$$
$$\begin{array}{r} 15 \\ -\ 9 \\ \hline \end{array}$$
$$\begin{array}{r} 14 \\ -\ 8 \\ \hline \end{array}$$
$$\begin{array}{r} 13 \\ -\ 7 \\ \hline \end{array}$$
$$\begin{array}{r} 14 \\ -\ 7 \\ \hline \end{array}$$
$$\begin{array}{r} 15 \\ -\ 6 \\ \hline \end{array}$$

4.
$$\begin{array}{r} 14 \\ -\ 9 \\ \hline \end{array}$$
$$\begin{array}{r} 12 \\ -\ 7 \\ \hline \end{array}$$
$$\begin{array}{r} 16 \\ -\ 7 \\ \hline \end{array}$$
$$\begin{array}{r} 14 \\ -\ 6 \\ \hline \end{array}$$
$$\begin{array}{r} 16 \\ -\ 9 \\ \hline \end{array}$$
$$\begin{array}{r} 11 \\ -\ 6 \\ \hline \end{array}$$

5.
$$\begin{array}{r} 15 \\ -\ 8 \\ \hline \end{array}$$
$$\begin{array}{r} 16 \\ -\ 8 \\ \hline \end{array}$$
$$\begin{array}{r} 13 \\ -\ 9 \\ \hline \end{array}$$
$$\begin{array}{r} 12 \\ -\ 3 \\ \hline \end{array}$$
$$\begin{array}{r} 15 \\ -\ 7 \\ \hline \end{array}$$
$$\begin{array}{r} 13 \\ -\ 5 \\ \hline \end{array}$$

Use with Lesson 11-4, text pages 421–422.

Name _____

Add. Write the related fact.

1.

8 +9	9 +8	9 +7	8 +7	9 +9

Subtract. Check by adding.

2.

15 − 8	7 +8	17 − 8	16 − 9	16 − 8

Write the fact family.

3. 14 8 6

___6__ + ___8__ = ____   ____ + ____ = ____

____ − ____ = ____   ____ − ____ = ____

4. 17 9 8

___9__ + ___8__ = ____   ____ + ____ = ____

____ − ____ = ____   ____ − ____ = ____

5. 18 9

____ + ____ = ____   ____ − ____ = ____

6. 6 12

____ + ____ = ____   ____ − ____ = ____

Use with Lessons II-5 and II-6, text pages 423–426.

Change the order of the addends.
Ring the group you add first.
Find the sum.

1. $5 + 7 + 5 =$ (5 + 5) + 7 = 17

2. $2 + 6 + 8 =$ ___ + ___ + ___ = ___

3. $4 + 8 + 4 =$ ___ + ___ + ___ = ___

4. $1 + 8 + 9 =$ ___ + ___ + ___ = ___

5. $6 + 7 + 4 =$ ___ + ___ + ___ = ___

6. $9 + 0 + 9 =$ ___ + ___ + ___ = ___

7. $3 + 5 + 7 =$ ___ + ___ + ___ = ___

Add up or down.

8.

3	2	2	1	4	8
7	5	8	9	6	7
+9	+5	+7	+9	+7	+3
19					

9.

3	6	3	7	4	8
4	6	8	7	4	1
+6	+3	+1	+2	+7	+5

Solve. Draw a line through the extra information.

1. I need 14 forks. ~~I have 14 knives.~~ I have 9 forks.
   How many more forks do I need?

   14 ⊖ 9 = ____

   I need ____ forks.

---

2. 7 cups are blue. 9 cups are green. 6 plates are blue.
   How many cups are there in all?

   ____ ◯ ____ = ____

   ____ cups in all

---

3. 8 people have punch. 4 people want juice.
   Ted gives 7 more people punch.
   How many have punch now?

   ____ ◯ ____ = ____

   ____ have punch.

---

4. 18 friends come to lunch. 8 friends are girls.
   9 friends have left. How many are still at lunch?

   ____ ◯ ____ = ____

   ____ are still at lunch.

Use with Lesson 11-8, text pages 429–430.

Name _____

Read → Draw → Think → Write

Shelly

1. Lynn got 7 ⬤.

   Shelly got 14 ⬤.

   How many more
   did Shelly get?

   Lynn

   add
   or
   (subtract)

   $\begin{array}{r} 1\,4 \\ -\,7 \\ \hline \end{array}$

   Shelly got ____ ⬤ more.

2. Ann put the ninth
   ⛵ on the shelf.

   Then she put 7 more
   ⛵ on the shelf.

   How many ⛵ are
   on the shelf?

   (add)
   or
   subtract

   $\begin{array}{r} \\ + \\ \hline \end{array}$

   There are ____ ⛵ in all.

3. Marta has 8 🐟.

   Karl has 9 🐟.

   How many 🐟 do

   Marta and Karl have?   Marta and Karl have ____ 🐟.

4. Brian made 15 🍪
   for the team. There
   are 6 🍪 left.

   How many 🍪 did
   the team eat?   The team ate ____ 🍪.

Use with Lesson 11-9, text pages 431–432.

141

Fun with Facts to 18

Name _____

First add or subtract.
Then color.

red — 6, 7, 8		blue — 15
brown — 9, 10		purple — 16
green — 11, 12		orange — 17
yellow — 13, 14		pink — 18

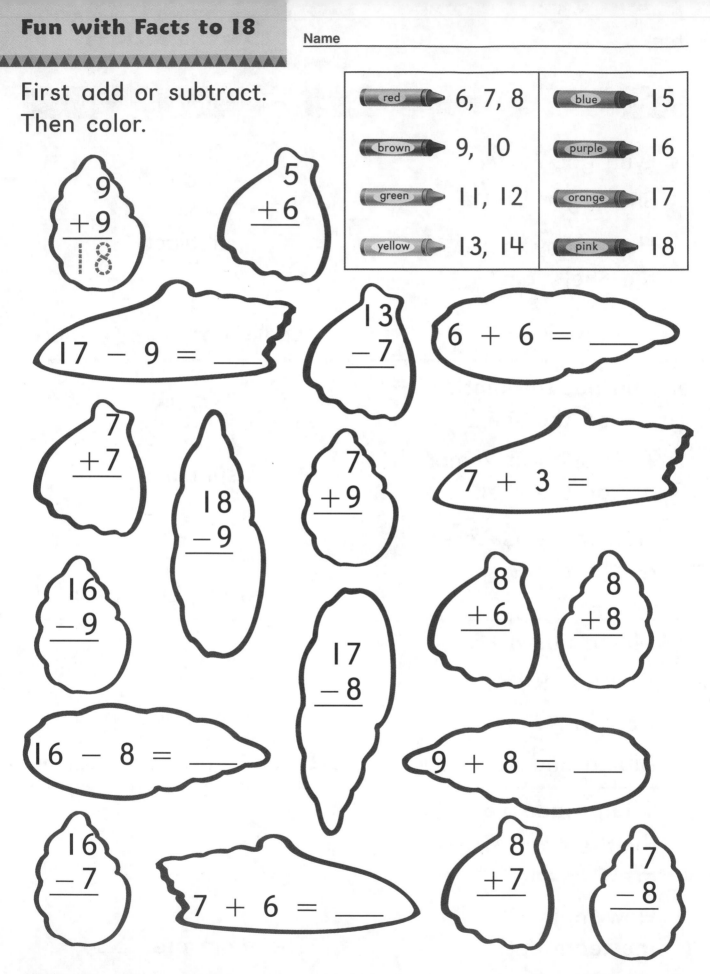

$$\begin{array}{r} 9 \\ +\,9 \\ \hline 18 \end{array}$$

$$\begin{array}{r} 5 \\ +\,6 \\ \hline \end{array}$$

$$17 - 9 = \underline{}$$

$$\begin{array}{r} 13 \\ -\,7 \\ \hline \end{array}$$

$$6 + 6 = \underline{}$$

$$\begin{array}{r} 7 \\ +\,7 \\ \hline \end{array}$$

$$\begin{array}{r} 18 \\ -\,9 \\ \hline \end{array}$$

$$\begin{array}{r} 7 \\ +\,9 \\ \hline \end{array}$$

$$7 + 3 = \underline{}$$

$$\begin{array}{r} 16 \\ -\,9 \\ \hline \end{array}$$

$$\begin{array}{r} 8 \\ +\,6 \\ \hline \end{array}$$

$$\begin{array}{r} 8 \\ +\,8 \\ \hline \end{array}$$

$$\begin{array}{r} 17 \\ -\,8 \\ \hline \end{array}$$

$$16 - 8 = \underline{}$$

$$9 + 8 = \underline{}$$

$$\begin{array}{r} 16 \\ -\,7 \\ \hline \end{array}$$

$$7 + 6 = \underline{}$$

$$\begin{array}{r} 8 \\ +\,7 \\ \hline \end{array}$$

$$\begin{array}{r} 17 \\ -\,8 \\ \hline \end{array}$$

Use after Chapter 11.

Name _____

Solve. Use a strategy you know.

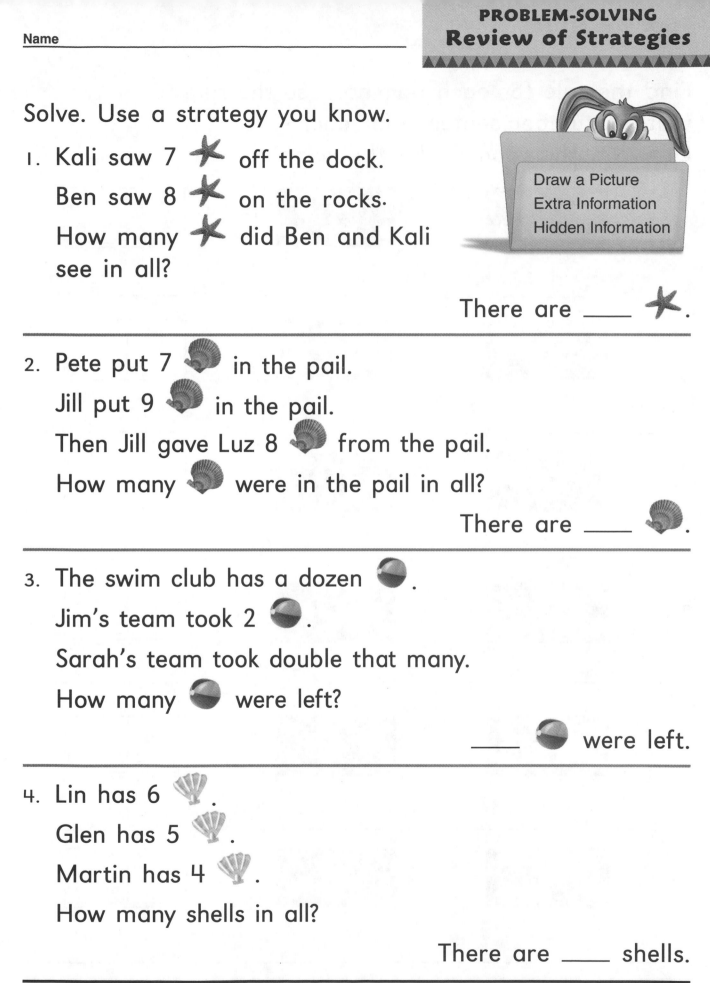

1. Kali saw 7 ⭐ off the dock.
   Ben saw 8 ⭐ on the rocks.
   How many ⭐ did Ben and Kali
   see in all?

   There are ____ ⭐.

2. Pete put 7 🐚 in the pail.
   Jill put 9 🐚 in the pail.
   Then Jill gave Luz 8 🐚 from the pail.
   How many 🐚 were in the pail in all?

   There are ____ 🐚.

3. The swim club has a dozen 🏐.
   Jim's team took 2 🏐.
   Sarah's team took double that many.
   How many 🏐 were left?

   ____ 🏐 were left.

4. Lin has 6 🐚.
   Glen has 5 🐚.
   Martin has 4 🐚.
   How many shells in all?

   There are ____ shells.

Draw a Picture
Extra Information
Hidden Information

Find the Rule

Find the rule for each domino. Use the rule to
write a number sentence for each.
Draw another domino for the same rule.

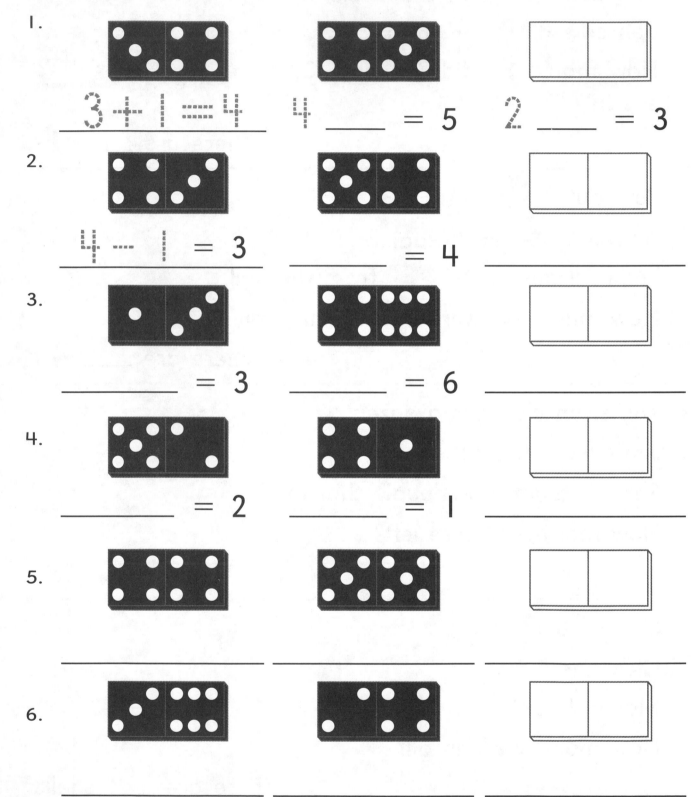

1.

$3 + 1 = 4$ 4 ____ $= 5$ 2 ____ $= 3$

2.

$4 - 1 = 3$ ____ $= 4$

3.

____ $= 3$ ____ $= 6$

4.

____ $= 2$ ____ $= 1$

5.

6.

Use with Lesson 12-1, text page 443.

Name _____

Fill in the ☐. Solve to check.

1.
$$13 - 7 = \boxed{4} + 2$$
$$\underline{6} = \underline{6}$$

2.
$$\boxed{} + 5 = 2 + 12$$
$$\underline{} = \underline{}$$

3.
$$7 + 8 = 10 + \boxed{}$$
$$\underline{} = \underline{}$$

4.
$$56 - 40 = \boxed{} + 8$$
$$\underline{} = \underline{}$$

5.
$$\boxed{} + 20 = 10 + 15$$
$$\underline{} = \underline{}$$

6.
$$26 - 20 = \boxed{} + 2$$
$$\underline{} = \underline{}$$

7.
$$\boxed{} - 10 = 6 + 4$$
$$\underline{} = \underline{}$$

8.
$$32 - \boxed{} = 6 + 6$$
$$\underline{} = \underline{}$$

9.
$$7 + 7 = \boxed{} + 4$$
$$\underline{} = \underline{}$$

10.
$$26 - 6 = 10 + \boxed{}$$
$$\underline{} = \underline{}$$

11.
$$9 + 9 = 8 + \boxed{}$$
$$\underline{} = \underline{}$$

12.
$$27 - \boxed{} = 8 + 9$$
$$\underline{} = \underline{}$$

Name _____

Fill in the missing + and − signs.

1.

$$12 \ominus 10 \oplus 2 = 4$$

2.

$$9 \bigcirc 5 \bigcirc 8 = 12$$

3.

$$3 \bigcirc 5 \bigcirc 9 = 17$$

4.

$$12 \bigcirc 5 \bigcirc 3 = 10$$

5.

$$8 \bigcirc 5 \bigcirc 3 = 6$$

6.

$$13 \bigcirc 9 \bigcirc 8 = 12$$

7.

$$7 \bigcirc 3 \bigcirc 2 = 12$$

8.

$$6 \bigcirc 6 \bigcirc 8 = 4$$

9.

$$12 \bigcirc 2 \bigcirc 3 = 7$$

10.

$$5 \bigcirc 5 \bigcirc 8 = 2$$

11.

$$11 \bigcirc 6 \bigcirc 2 = 3$$

12.

$$7 \bigcirc 3 \bigcirc 10 = 20$$

Use with Lesson 12-3, text page 445.

Name _____

Find the missing number.

1. ◯ = 5   ▢ + ◯ = 9 $4 + 5 = 9$

   ▢ = ?   ▢ + 5 = 9   ▢ = 4

2. △ = 4   ▭ − △ = 8   _____

   ▭ = ?   ▭ − __ = __   ▭ = __

3. ▢ = 13   ▢ − ⬭ = 6   _____

   ⬭ = ?   __ − ⬭ = __   ⬭ = __

4. ▱ = 6   ▱ + ◯ = 11   _____

   ◯ = ?   __ + ◯ = __   ◯ = __

5. ⬠ = 10   ⬠ + 30 + ▢ = 48   _____

   ▢ = ? __ + __ + ▢ = __   ▢ = __

Regrouping Money

Name _____

Find the sum. Regroup where needed.

1.
$$73¢ + 19¢ = 92¢$$
$$28¢ + 56¢$$
$$19¢ + 3¢$$
$$24¢ + 9¢$$
$$28¢ + 47¢$$

2.
$$24¢ + 68¢$$
$$48¢ + 22¢$$
$$53¢ + 6¢$$
$$29¢ + 18¢$$
$$16¢ + 17¢$$

3.
$$46¢ + 9¢$$
$$30¢ + 4¢$$
$$15¢ + 28¢$$
$$29¢ + 33¢$$
$$75¢ + 5¢$$

4.
$$51¢ + 9¢$$
$$18¢ + 14¢$$
$$53¢ + 44¢$$
$$25¢ + 49¢$$
$$18¢ + 29¢$$

Find the difference. Regroup where needed.

5.
$$40¢ - 17¢ = 23¢$$
$$77¢ - 44¢$$
$$35¢ - 8¢$$
$$68¢ - 34¢$$
$$50¢ - 15¢$$

6.
$$85¢ - 23¢$$
$$42¢ - 17¢$$
$$60¢ - 25¢$$
$$55¢ - 13¢$$
$$46¢ - 7¢$$

Use with Lesson 12-5, text pages 447–448.

Add.

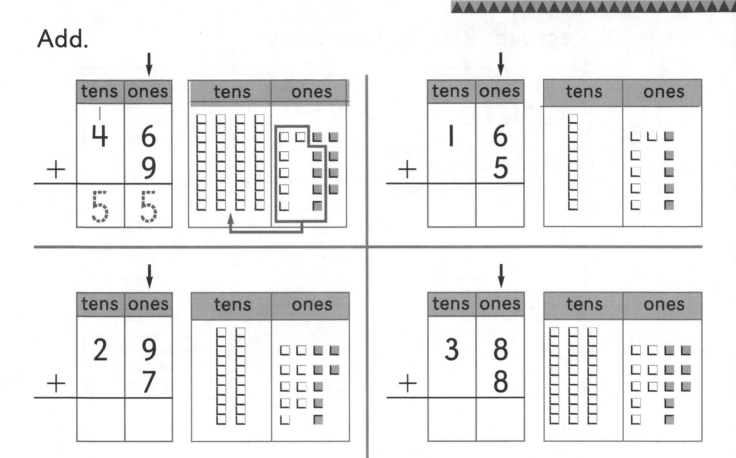

	tens	ones
	4	6
+		9
	5	5

	tens	ones
	1	6
+		5

	tens	ones
	2	9
+		7

	tens	ones
	3	8
+		8

Find the sum.

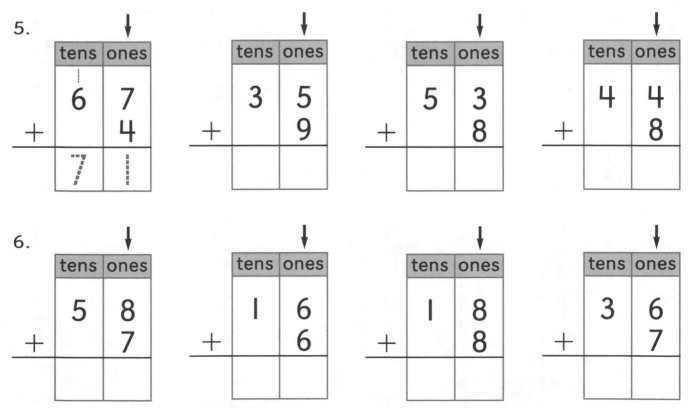

5.

	tens	ones
	6	7
+		4
	7	1

	tens	ones
	3	5
+		9

	tens	ones
	5	3
+		8

	tens	ones
	4	4
+		8

6.

	tens	ones
	5	8
+		7

	tens	ones
	1	6
+		6

	tens	ones
	1	8
+		8

	tens	ones
	3	6
+		7

Use with Lesson 12-6, text pages 449–450.

Subtract. Regroup. Ring to take away.

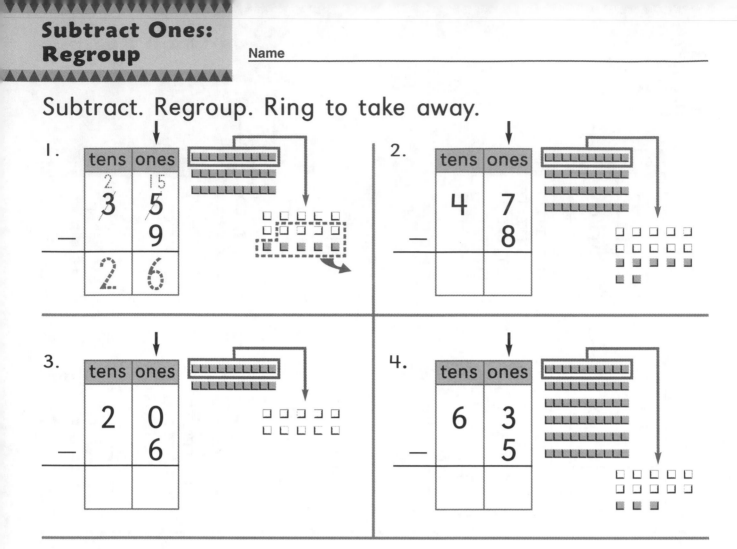

1.

tens	ones
2	15
3̸	5̸
−	9
2	6

2.

tens	ones
4	7
−	8

3.

tens	ones
2	0
−	6

4.

tens	ones
6	3
−	5

Find the difference.

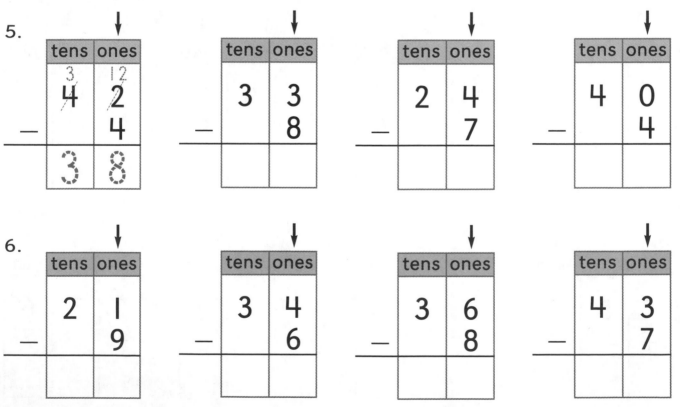

5.

tens	ones
3	12
4̸	2̸
−	4
3	8

tens	ones
3	3
−	8

tens	ones
2	4
−	7

tens	ones
4	0
−	4

6.

tens	ones
2	1
−	9

tens	ones
3	4
−	6

tens	ones
3	6
−	8

tens	ones
4	3
−	7

Use with Lesson 12-7, text pages 451–452.

Name

Find the sum.

1.
```
   21        23        54        31        35
   16        52        11        45        13
 + 31      + 12      + 13      + 13      + 10
  ----      ----      ----      ----      ----
   68
```

2.
```
   64        32        50        41        16
   22        31        13        25        60
 + 13      + 14      + 31      + 21      + 12
  ----      ----      ----      ----      ----
```

Use the numbers in the boxes.

Write the missing addends.

| 22 | | 13 |

| 34 | | 12 |

3.
$24 + \underline{\quad} + \underline{\quad} = 59$

4.
$32 + \underline{\quad} + \underline{\quad} = 78$

5.
$32 + \underline{\quad} + \underline{\quad} = 88$

6.
$33 + \underline{\quad} + \underline{\quad} = 79$

7.
$32 + \underline{\quad} + \underline{\quad} = 57$

8.
$20 + \underline{\quad} + \underline{\quad} = 67$

Use with Lesson 12-8, text pages 453–454.

Hundred, Tens, Ones

Name _____

Write the place value and the number.

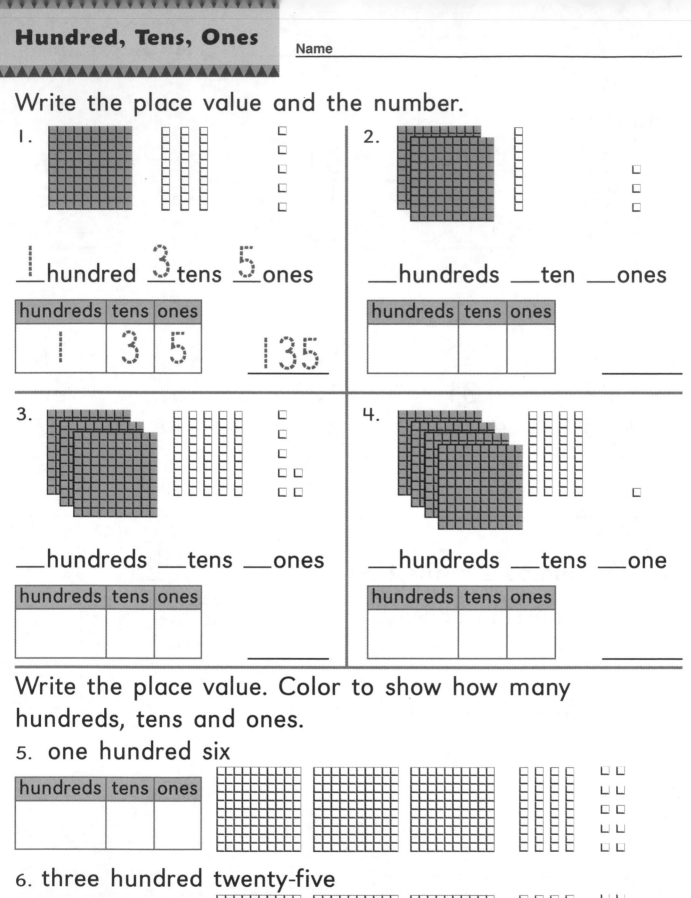

1. _1_ hundred _3_ tens _5_ ones

hundreds	tens	ones
1	3	5

_135_

2. ___ hundreds ___ ten ___ ones

hundreds	tens	ones

_____

3. ___ hundreds ___ tens ___ ones

hundreds	tens	ones

_____

4. ___ hundreds ___ tens ___ one

hundreds	tens	ones

_____

Write the place value. Color to show how many hundreds, tens and ones.

5. one hundred six

hundreds	tens	ones

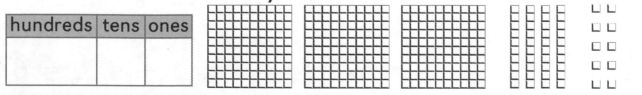

6. three hundred twenty-five

hundreds	tens	ones

Name _____

Find the sum. Use models to help.

1. $342 + 134$

h	t	o
3	4	2
+ 1	3	4
4	7	6

2. $121 + 450$

h	t	o
+		

3. $433 + 216$

h	t	o
+		

Find the difference. Use models to help.

4. $695 - 254$

h	t	o
6	9	5
− 2	5	4
4	4	1

5. $575 - 341$

h	t	o
−		

6. $329 - 113$

h	t	o
−		

Add or subtract. Watch the + or − signs.

7.
$$527 - 306 = 221$$
$$417 + 581$$
$$164 + 211$$
$$354 - 103$$
$$425 - 120$$

8.
$$653 - 302$$
$$174 + 112$$
$$986 - 572$$
$$205 + 680$$
$$375 - 264$$

1.

Read · Think · Write · Check

Rides cost 10¢ each.
How many rides can I take for 50¢?

Complete the table. Look for a pattern.

money	10¢	20¢	30¢	40¢	50¢
rides	1	2			

_5_ rides for 50¢.

2.

Read · Think · Write · Check

A sticker costs 5¢.
How much do I pay for 6 stickers?

Complete the table. Look for a pattern.

stickers	1	2				
cost	5¢	10¢				

I pay _____ for 6 stickers.

3.

Read · Think · Write · Check

There are 16 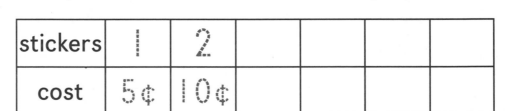. Two go in each bumper car. How many cars have ?

Make a table.

2							
1							

____ cars have .

Use with Lesson 12-11, text pages 459–460.

Name _____

Match the numbers with their models.

1. I am a 3-digit number.
   I have less than 3 hundreds.
   I have no tens.
   What number am I?

   I am ___102___.

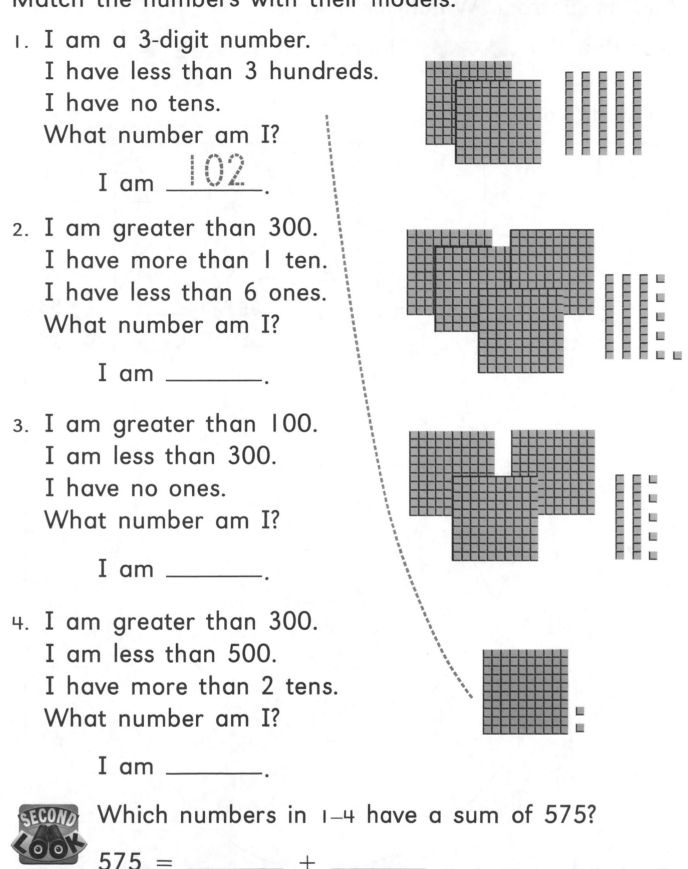

2. I am greater than 300.
   I have more than 1 ten.
   I have less than 6 ones.
   What number am I?

   I am _____.

3. I am greater than 100.
   I am less than 300.
   I have no ones.
   What number am I?

   I am _____.

4. I am greater than 300.
   I am less than 500.
   I have more than 2 tens.
   What number am I?

   I am _____.

Which numbers in 1–4 have a sum of 575?

575 = _____ + _____

Add or subtract. Watch for + and −.

Less than 50 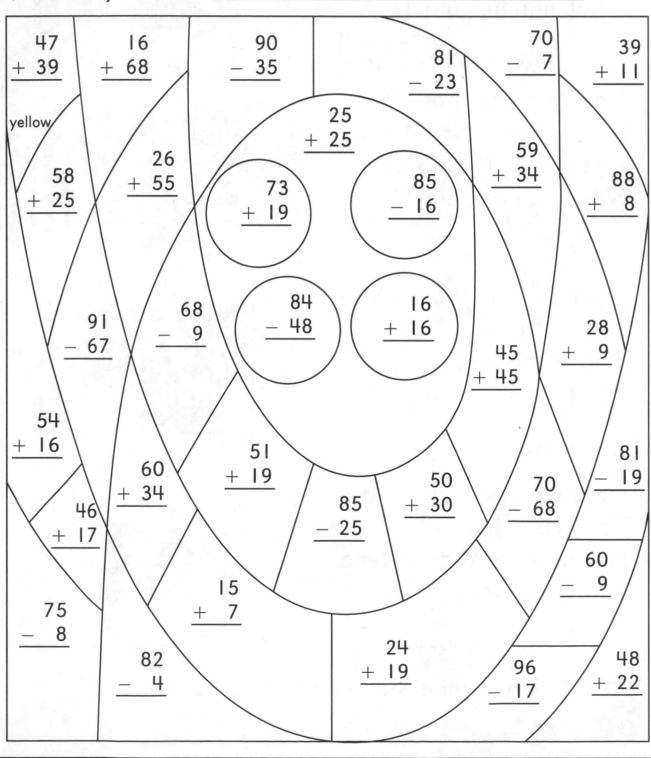 red . More than 50 yellow .

What do you see? _____

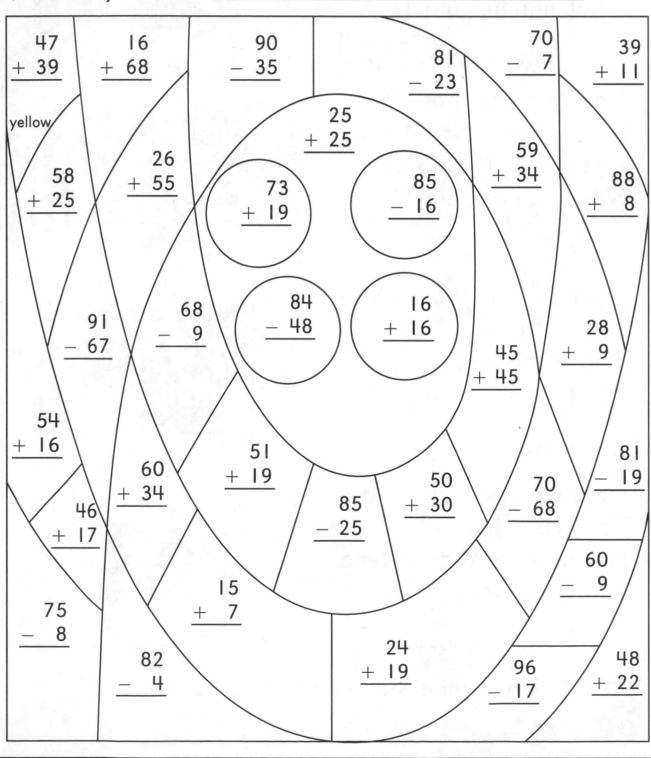

Use after Chapter 12.